AF748762

Anticipation, Sustainability, Futures and Human Extinction

This book considers the philosophical underpinnings, policy foundations, institutional innovations, and deep cultural changes needed to ensure that humanity has the best chance of surviving and flourishing into the very distant future.

Anticipation of threats to the sustainability of human civilization needs to encompass time periods that span not just decades but millennia. All existential risks need to be jointly assessed, as opposed to addressing risks such as climate change and pandemics separately. Exploring the potential events that are likely to cause the biggest risks as well as asking why we should even desire to thrive into the distant future, this work looks at the 'biggest picture possible' in order to argue that futures-oriented decision-making ought to be a permanent aspect of human society and futures-oriented policy making must take precedent over the day-to-day policy making of current generations in times of great peril. The book concludes with a discourse on the truly fundamental bottom-up changes needed in our personal psychologies and culture to support these top-down recommendations.

This book will be of great interest to philosophers, policy analysts, political scientists, economists, psychologists, planners, and theologians.

Bruce E. Tonn is the President of Three^3, a scientific and educational non-profit organization, whose mission is to foster equitable sustainable futures. He was previously Professor of Public Policy at The University of Tennessee, Knoxville, USA.

Routledge Research in Anticipation and Futures
Series editors: Johan Siebers and Keri Facer

The Promise of Nostalgia
Reminiscence, Longing and Hope in Contemporary American Culture
Nicola Sayers

Indigenous Futures and Learnings Taking Place
Edited by Ligia (Licho) López López and Gioconda Coello

Sustainable and Democratic Education
Opening Spaces for Complexity, Subjectivity and the Future
Sarah Chave

A Creative Philosophy of Anticipation
Futures in the Gaps of the Present
Edited by Jamie Brassett and John O'Reilly

Anticipation, Sustainability, Futures and Human Extinction
Ensuring Humanity's Journey into the Distant Future
Bruce E. Tonn

For more information about this series, please visit: www.routledge.com/Routledge-Research-in-Anticipation-and-Futures/book-series/RRAF

Anticipation, Sustainability, Futures and Human Extinction

Ensuring Humanity's Journey into the Distant Future

Bruce E. Tonn

Routledge
Taylor & Francis Group
LONDON AND NEW YORK

First published 2021
by Routledge
2 Park Square, Milton Park, Abingdon, Oxon OX14 4RN

and by Routledge
52 Vanderbilt Avenue, New York, NY 10017

Routledge is an imprint of the Taylor & Francis Group, an informa business

© 2021 Bruce E. Tonn

The right of Bruce E. Tonn to be identified as author of this work has been asserted by him in accordance with sections 77 and 78 of the Copyright, Designs and Patents Act 1988.

All rights reserved. No part of this book may be reprinted or reproduced or utilised in any form or by any electronic, mechanical, or other means, now known or hereafter invented, including photocopying and recording, or in any information storage or retrieval system, without permission in writing from the publishers.

Trademark notice: Product or corporate names may be trademarks or registered trademarks, and are used only for identification and explanation without intent to infringe.

British Library Cataloguing-in-Publication Data
A catalogue record for this book is available from the British Library

Library of Congress Cataloging-in-Publication Data
Names: Tonn, Bruce E. (Bruce Edward), 1955– author.
Title: Anticipation, sustainability, futures and human extinction : ensuring humanity's journey into the distant future / Bruce E. Tonn.
Description: Abingdon, Oxon ; New York, NY : Routledge, 2020. | Includes bibliographical references and index.
Identifiers: LCCN 2020052911 (print) | LCCN 2020052912 (ebook)
Subjects: LCSH: Sustainability—Philosophy. | Sustainability—Government policy. | Human ecology. | Human beings—Extinction | Human beings—Forecasting. | Ecological forecasting | Expectation (Philosophy)
Classification: LCC GE196 .T66 2020 (print) | LCC GE196 (ebook) | DDC 304.2—dc23
LC record available at https://lccn.loc.gov/2020052911
LC ebook record available at https://lccn.loc.gov/2020052912

ISBN: 978-0-367-42913-3 (hbk)
ISBN: 978-0-367-76757-0 (pbk)
ISBN: 978-1-003-00010-5 (ebk)

Typeset in Baskerville
by Apex CoVantage, LLC

This book is dedicated to my children, their children, and all of humanity's children yet to come.

Contents

Figures

Tables

Acknowledgments

I acknowledge and appreciate the support I have received from many institutions and people. I am grateful that I had the privilege of attending three excellent universities, Stanford, Harvard, and Northwestern. I want to thank George Peterson for his support in guiding me through the Ph.D. program in urban and regional planning at Northwestern University.

I want to thank Jim Dator and Alan Tough for welcoming me to the field of futures studies at the Kyoto Futures Forum. I also want to thank Wendell Bell and Richard Slaughter for their compendiums about the field. I am greatly appreciative of the opportunities that Ziauddin Sardar and Ted Fuller have given me to contribute to the editorial process of the journal *Futures.* I wish to thank my colleagues and students at the University of Tennessee, Knoxville, who were enthusiastic about my futures-orientation to governance, bioregional planning, and energy and environmental policy.

I want to thank my friends and colleagues who provided reviews of drafts of this book, including Dori Stiefel, Paula Shaffer, Ted Fuller, Dave Feldman, Seth Baum, Diana Tonn, Erin Rose, Beth Hawkins, Michaela Marincic, A. J. Kukay, Loes Damhof, Annette Gardner, and Don MacGregor. I also wish to thank Clay Biden, who conducted background research for the earlier chapters of the book. I want to say a special thanks to Dori Stiefel, who has been a patient co-author of many of the papers whose substance made it into this book. I also want to express my appreciation for my friendships with Don MacGregor, Dave Feldman, Woody Dowling, Fred Conrad, and William Sims Bainbridge, who are always ready to tackle any sort of intellectual questions.

I want to thank my close friends and colleagues Erin Rose and Beth Hawkins. We founded our non-profit research organization, Three^3 (pronounced ThreeCubed) in 2013, with a mission *to foster equitable, sustainable futures.* Thus, both intergenerational and intra-generational equity are important in our research. Our funded research focuses on the intersection of energy efficiency, housing, poverty, and climate change. Therefore, this book fits perfectly within our mission.

Lastly, I want to thank dear friends and family. I want to thank David Telleen-Lawton and Christopher Thomsen for being lifelong friends. I want to thank my brother, David Evans, for being a true inspiration. I want to thank my wife, Diana, who has been my constant companion since we met on a train going from Paris to Stuttgart on our way to Stanford's overseas program in Beutelsbach, Germany. I want to thank her for her support for my futures research, and of course, as being my better half in raising our three children, Jenna, Chris, and Shara. I want to also thank Robert and Jeanne Bettencourt, Diana's parents, for their love and support over the years.

1 Introduction

Humanity will not survive into the distant future merely by chance. There are simply too many catastrophic risks threatening our existence. Many existential risks, such as climate change and loss of biodiversity, have been years in the making and will require years of concerted effort to reduce. Others risks, such as from asteroids and pandemics,[1] can strike relatively quickly and require years of concerted preparation to protect ourselves and generations of our children. We need to anticipate these and other risks to humanity and take appropriate proactive actions to reduce the risks. We need to improve our decision-making to increase the probability of our survival.

That humanity struggles with being futures-oriented is too kind a portrayal because pervasive myopia characterizes every aspect of human society. Politicians' attentions are focused on the next election and on managing the 24-hour news cycle. Companies fixate on quarterly and annual earnings reports. Traders of financial securities employ software systems that reduce decision-making timescales to nanoseconds. The intellectual underpinnings of economic theory and public administration heavily discount the future[2] and settle for 'muddling through',[3] respectively. Over half of the constitutions of nations around the world fail to use any derivation of the words, 'future' or 'posterity'.[4] Religious beliefs that embrace determinism and fatalism implicitly communicate that our futures are beyond human control. Individualism enshrines narcissism and diverts attention from the shared human projects of anticipation, sustainability, and futures-oriented thinking. Individuals find thinking about futures anxiety provoking and future times unimaginable even a few years into the future.[5]

One can draw at least two conclusions from this characterization. The first is that the pervasive myopia that characterizes every aspect of human society implies that we naively and stubbornly believe, to the ultimate harm of our future generations, that we actually can survive any gauntlet of existential risks without investing time and effort in foresight and anticipation. The second is that we have not given our futures and the possibility of our own extinction much thought because we do not know how to go about it in any way, shape, or form.

The second position is the more optimistic one to take and is the motivation for this book. The main objective here is to provide a comprehensive *blueprint* for humanity to use to safeguard our journey into the distant future. This blueprint synthesizes a deep futures-perspective with philosophical underpinnings and institutional innovations to build an action-oriented and futures-oriented policy making system designed to reduce the probability of global catastrophes and human extinction.

The aims of this book are to explain what it means to be futures-oriented; discuss why we should be futures-oriented; propose *perpetual obligations* that if met by current generations will help ensure humanity survives into the distant future; propose metrics to determine whether we are meeting our perpetual obligations; discuss how we can anticipate whether we are moving toward meeting or not meeting the perpetual obligations; propose actions we as a global society must take if we are not meeting the perpetual obligations; and suggest indicative changes in the foundations of political systems, economies, and cultures to build more futures-oriented societies.

This book is unique in that it ties all of these components together. Books on sustainability certainly are futures-oriented but leave time frames of analysis implicit or completely unstated and rarely take on the topics of catastrophic risks and human extinction. Books that anticipate such risks typically do not address sustainability issues nor take on the challenge of explaining explicitly why humanity should strive to survive into the deep future. Books that address philosophical issues about why we should care about future generations rarely make any links with practical policy making frameworks. The many books on climate change also largely ignore philosophical issues and, by definition, ignore the many other existential risks to humanity.

Thus, the main theme running through this book is 'the biggest picture possible'. In other words, this book shows why and how *all* risks to humanity ought to be addressed concurrently, not just one or the other separately. This book takes a vast perspective on sustainability, that life on earth needs to be sustained for tens of millions of years, not just the next century. Futures-oriented decision-making ought to be a permanent aspect of human society and futures-oriented policy making must take precedent over the day-to-day policy making of current generations in times of great existential peril. Anticipating risks to humanity must be constant, rigorous, and time expansive. Adopting the biggest picture possible perspectives addressed in this book can be seen as an essential next step in the progress of human civilization.

Definition and linking of terms

Before diving into the main parts of this book, let's first take a minute to define and link together the four main concepts of this book: sustainability; anticipation; futures; and human extinction.

Sustainability and sustainable development are often interchangeable terms. The Brundtland Commission famously defines the latter as "development that meets the needs of the present without compromising the ability of future generations to meet their own needs."[6] Herman Daly states that sustainability should hew to these three principles: non-renewable resources should not be depleted at rates higher than the development rate of renewable resources; renewable resources should not be exploited at a rate higher than their regeneration level; and the absorption and regeneration capacity of the natural environment should not be exceeded.[7] Donella Meadows states:

> Our rational minds tell us that a sustainable world has to be one in which renewable resources are used no faster than they regenerate; in which pollution is emitted no faster than it can be recycled or rendered harmless; in which population is at least stable, maybe decreasing; in which prices internalize all costs; in which there is no hunger or poverty; in which there is true enduring democracy. But what else?[8]

Two important observations can be made about sustainability given these definitions. First, our global society writ large is highly unsustainable, meaning that we simply cannot continue to behave and consume as we are now into the indefinite future. We are drawing down our fossil fuel resources much faster than we are developing renewable energy resources.[9] Our demands for clean water already exceed supplies. Negative trends with respect to species extinction, soil quality, desertification, and hazardous wastes are accelerating. This interpretation of sustainability is key to this book and has also been taken up by governments, companies, communities, and individuals across the world.

Also important to this book is the observation that these and most other definitions of sustainability are inexplicably vague with respect to time. None of them answers this question: sustainable by when and for how long into the future? The biggest picture possible interpretation of sustainability requires the earth to sustain life for billions of humans for many millions of years. This expanded perspective extends concerns about resource use to encompass the stewardship of global ecosystems so they will be able to sustain life into the far future.

Anticipation is also a multifaceted concept. It can indicate the excitement someone has about an upcoming event, such as a birthday or concert. More technically, "An anticipatory system is a system containing a predictive model of itself and/or its environment, which allows it to change state at an instant in accord with the model's predictions pertaining to a later instant."[10] More simply put, anticipated future states of the world may require present changes in behaviors, as appropriate.[11] Anticipation is a concept that shares aspects with other concepts such as prediction and forecast in that it focuses on identifying future states of the world that may be of concern. Anticipation is a more forgiving concept in that it does not require the trappings of technical precision usually associated with the other two concepts, precision that is often unsupportable and pretentious and that then, perversely, damages wise proactive decision-making.

With respect to sustainability, anticipation takes on a more sober personality. It is the act of anticipating threats to the sustainability of key systems and systems of systems and changing behaviors today so we can anticipate more sustainable futures. For example, we can anticipate the exhaustion of fossil fuels and subsequent collapse of modern economies. We can anticipate economic disruptions and social troubles that could lead to needless wars and uncontrolled pandemics. The challenge is to change behaviors expeditiously and justly so that we can anticipate more favorable futures.

Thus, anticipation is closely linked to proactive behavior. Unfortunately, history has demonstrated time and again that proactive behavior by our political institutions, at the very least, is the exception, not the norm. Individuals, organizations, and governments constantly react to crises that could have been prevented or ameliorated if they had spent any time anticipating such events. Only infrequently are proactive policies adopted and oftentimes these are only adopted after repeated devastating calamities. In the United States, for example, homes and communities must suffer from urban flooding numerous times before proactive policies are even considered.[12] Methods drawn from futures studies and other fields can improve our anticipatory prowess.[13]

Futures is a term that has an interesting backstory.[14] The field of futures studies was borne during the 1950s and 1960s by work done by RAND for the U.S. Department of Defense.[15] Western and Eastern bloc countries both were interested in futures-thinking, the West primarily with respect to environmental and technological futures and the East with respect to the inexorable march of time toward communism. The latter, by dogma and convention, believed in one and only future, *the* future. The West, in contrast, believed that the future could take on many forms, largely dependent upon decisions made by policy makers and others. A major debate raged between the two sides, with the West eventually adopting the plural, *futures*, to define the field of futures studies. De Jourvenl used the plural 'futures' rather than the singular 'future' in his influential *The Art of Conjecture*, feeling that this represented the best perspective in which to conduct futures studies. This preference was then embraced broadly by the West.[16]

The field was popularized by books such as *Future Shock*[17] and *Megatrends*,[18] which were about major trends that could impact and even substantially disrupt society. Al Gore's book, *The Future*,[19] carries on the lineage of trends assessment. The book *Art of the Long View* is widely used by practitioners who conduct trend analyses and write scenarios.[20] The academic foundations of future studies were established by the important texts written by Wendell Bell[21] and edited by Richard Slaughter.[22] These authors addressed the epistemological foundations of futures studies. Numerous authors have further developed our knowledge base about scenario writing.[23]

My own introduction to the field of futures studies was serendipitous. I was hired by Oak Ridge National Laboratory in 1983 after completing my Ph.D. In those days, my research tasks frequently led me to ORNL's library to consult hardcopies of journals. While browsing the shelves of resources, I happened across the journal *Futures.* I had a few extra minutes to thumb through several volumes. It was immediately clear to me that the articles in *Futures* filled a gap in my education, complemented my planning sensibilities, and also fit well with many of the major issues being dealt with by scientists at ORNL at the time, including the safe disposal of nuclear wastes and the energy crisis. The first papers about the dangers of global warming, as climate change was referred to back then, were also being published. Thus, my interest in futures studies was born.[24]

Over the years, the term *futures* has come to connote to me both concerns about the future state of the world and freedom for human agency to shape the future, to imagine what the future could look like. Futures studies has helped to develop the literature as to why we should care about future generations and what are our obligations to future generations.[25] Most importantly, futures-thinking compels policy making in longer time frames, not months or years but decades, centuries, and even millennia or longer.[26]

Similar to sustainability and anticipation, it cannot be argued that society is futures-oriented, as noted earlier. It is rare for policy debates to be set in terms of decades or for policy to be made with respect to obligations to future generations. To our extreme detriment, virtually no aspect of human society is truly futures-oriented.

Human extinction is a straightforward concept. It means that at some point in the future, humans could no longer exist. We will have gone extinct like the dinosaurs, woolly mammoths, and dodos. It could also mean that humans evolve into post-humans through natural evolution or through advances of genetic engineering, should these techniques become accepted. Serious concerns about human extinction have been around since the dawn of the age of nuclear weapons, though one could argue that the concern could date back to the early 1800s.[27] Numerous other existential risks to humanity now join nuclear weapons, including climate change,[28] pandemics, asteroids, and super artificial intelligences that may have malevolent intents toward humanity.[29]

Let's now place and link these four terms in the largest picture possible. If we are to understand sustainability in the context of futures-thinking, then the goal of sustainability efforts is to sustain human life on planet Earth for tens of thousands and even hundreds of millions of years. If we are to understand anticipation in the context of human extinction, then the goal is to anticipate all threats to human extinction together, not only the most obvious, such as climate change and pandemics, separately.

Extending sustainability thinking to span millions of years and anticipating all risks to human extinction are difficult endeavors. The reason to take on these daunting tasks is to inform today's decision makers with the expectation that they will take actions to reduce the current and anticipated threats of global catastrophes and human extinction. Here we come to another central reason for this book; addressing how humanity should go about making such decisions.

The barriers to extensive and effective forward thinking and proactive action to protect humanity's future are steep. Here are some of the barriers addressed in this book: we fundamentally have little practice in or capacity for futures-thinking; we lack an appreciation of the full range of existential risks facing humanity; we lack a sound philosophical framework to explain why we should even care about surviving into the deep future; we lack a set of policy-oriented metrics (i.e., obligations to future generations) that if met can help ensure we will survive into the deep future; we lack a framework for humanity-wide actions to be taken if the obligations are not being met; we do not know how to estimate the probability of human extinction; we do not have institutions that can judge whether actions should be taken and if so, what level of actions; and, of course, our day-to-day psychologies, cultures, religions, and economics are nowhere suited for futures-thinking on this scale and scope.

This book takes on all of the barriers just stated. It draws upon the rich literature of the discipline of futures studies for its inspiration but is unique in that it ties together important concepts in the futures literature, such as why we should care about future generations and what are our obligations to future generations, to sustainability and anticipation within the context of real public policy decision-making. As such this book also draws upon and contributes to numerous other disciplines, from philosophy to political science, public policy analysis to decision theory.

Chapters overview

Chapter 2 is designed to nurture the futures-oriented perspective needed to lead the reader through the balance of the book. The chapter takes a serendipitous tour of anticipatable events though a time span from 50 years from now to the time when planet Earth becomes uninhabitable because of the death of our sun in about a billion years. The tour includes destinations such as the exhaustion of fossil fuels and the time when most of our current generations' nuclear wastes become harmless. The chapter also touches upon other anticipatable events, such as changes in human demographics and self-identity.

Chapter 3 takes a deep dive into the wide range of risks to our extinction. Added to the risks already mentioned, such as climate change, pandemics, and nuclear war are many less well-publicized risks, such as super-volcanoes and collisions with near-earth objects. The extinction-level risks are grouped into nine categories based on source (e.g., anthropogenic, extraterrestrial) and nature (e.g., current, emerging). The chapter cites many who believe that the risks of human extinction are already extraordinarily high, not only because the vast majority of earth's species have in fact gone extinct but also because the seemingly lack of intelligent life in the universe suggests that civilizations cannot survive their technological advancements. It is noted that for many extinction-level events of anthropogenic origin, the lead times needed to deal with these risks greatly exceed their expected time of occurrence. This chapter concludes with two short human extinction scenarios that are based on untoward consequences of emerging technologies.

The message conveyed by Chapter 4 is that the actual probability of human extinction is highly dependent upon our own behavior. We can lower the probability by anticipating risks, and by developing and implanting plans and policies to mitigate and ameliorate the risks. To improve our anticipatory prowess, we should (1) develop and assess scenarios that result in human extinction (Singular Chains of Events Scenarios); (2) come to grips with the unanticipated unintended consequences of purposive action (which includes emerging technologies); and (3) understand the importance of a rigorous and continuous program to deal with unknown unknowns. Implementation of a Whole Earth Monitoring System would provide us with the data needed to better assess risks. The chapter contains a preliminary scorecard about how well humanity is dealing with the nine categories of existential risks. At the end of the chapter is an example of a Singular Chain of Events Scenario.

Chapter 5 takes on this question: why should we care about future generations? Most people who worry about sustainability and climate change take it as a given that, of course, we should care about future generations but never really put forth an explicit, philosophically based argument about why we should care. This chapter reviews the literature related to why we should care, including stances taken by some that current generations have no obligations to future generations. This chapter disentangles the notions about why we should care about future generations from our obligations to future generations, which is the subject of Chapter 6. This chapter concludes that current generations ought to be futures-oriented as requirement of ethical living, not necessarily only framed with respect to future generations. In this sense, caring about future generations means we also care not only about the journey through time and space, but we are obsessive with protecting the rights and freedoms of future generations to also participate in *the journey*.

Chapter 6 reframes obligations to future generations as *perpetual obligations* that every current generation ought to strive to meet. A list of twelve perpetual obligations is presented. Eight of these are preventative, for example, we should prevent the extinction of humans; we should protect the essence of nature. Four are prescriptive, for example, we should always conduct research to learn more about how to survive into the future regardless of currently anticipated risks to humanity. The dozen perpetual obligations build upon previously proposed obligations, such as found in the United Nations Declaration of Human Rights, but the set is more comprehensive and, in a sense, updated to reflect the state of humanity in the early parts of the twenty-first century.

Several additional types of scenarios are introduced in Chapter 6 to complement Singular Chains of Events Scenarios. These include earth life extinction scenarios, global catastrophe scenarios, muddling through scenarios, and Transformative Scenarios (TS), an example of which is presented at the end of Chapter 9. Special attention is paid to determining what might be an acceptable risk threshold of human extinction. A scorecard of how humanity might be meeting each perpetual obligation during this period of time is included at the end of this chapter.

Chapter 7 dives into the nitty-gritty of policy making with respect to the twelve obligations. First, metrics are presented that can be used to help determine whether humanity is meeting the prescriptions inherent in each of the twelve perpetual obligations. Second, approaches to operationalizing each metric are proposed. The chapter sets out a framework for different levels of action depending on how far humanity is missing the mark with respect to its perpetual obligations, based on our performance on metrics. This framework has six levels, which run from Do Nothing to Manhattan Project to War Footing levels of effort, the latter reserved for situations where risks to human extinction are extraordinarily high and imminent. Where we are with respect to the framework for each perpetual obligation is addressed and assessed.

Chapter 7 also begins the work of tackling how to move from humanity-level actions to goals and actions that are more granular. First, the notion of 1000-year planning is introduced. Then, the discussion moves from actions linked to meeting perpetual obligations to more near-term global goals, as represented by the United Nations' Sustainable Development Goals.[30] Lastly, the chapter addresses how decision-making at the national and sub-national levels ought to synchronize with meeting perpetual obligations. A second humanity-wide system is proposed, called the Global Anticipation and Decision Support System (GADSS), to help decision makers understand the problems they and humanity need to address and also to help them anticipate what other decision makers may do in response.

Chapter 8 addresses the question of who has oversight responsibilities with respect to meeting perpetual obligations. More specifically, the chapter tackles who is (1) responsible for specifying and measuring the metrics, (2) determining which obligations are being met or not and if not, missing by what amount, and (3) determining and managing humanity's actions to meeting the spirit of the obligations. This chapter includes assessments of how national constitutions overwhelmingly do not address futures issues nor explicitly assign responsibilities to specific institutions to meet our obligations to future generations. In response, a set of five types of recommended amendments to national constitutions is presented: human rights commissions; ombudspersons; voting schemes that could be adopted by legislatures; assigning responsibilities to advocate for future generations to highly visible and influential existing institutions; and doing the same but to brand-new institutions. Examples of each are presented. Then the chapter moves on to various international organizational solutions, including one that would be modeled on the International Panel on Climate Change but with a focus on Perpetual Obligations. The chapter concludes with a discussion for the need for a small number of perpetual specialized institutions, such as one to act as a steward of the world's nuclear wastes.

Up to this point, the ideas presented in the book have a strong top-down essence. Chapter 9 argues that these ideas will only be acceptable and implementable if there is also a corresponding bottom-up effort as well. This is because virtually no aspects of our societies are futures-oriented. This chapter contains short discourses on politics, economics, culture, religion, and human psychology. Each discussion provides a small number of indicative ideas and suggestions for how that particular component could become more futures-oriented. The chapter also speculates that a convergence of many substantial changes to social-psychology and how society is administered to foster authentic futures-orientation could, over time, take human civilization through a Singularity of socio-economic change.[31] The chapter concludes with an example of an extensive Transformative Scenario.

Chapter 10 argues that humanity really does need to adopt the 'biggest picture possible' mentality to make effective and long-lasting progress toward the goal of ensuring that humanity survives into the very distant future. The book presents a blueprint. From an academic and research viewpoint, the book represents a goldmine of future research possibilities, as it is acknowledged that every idea, framework, proposal, and suggestion is open for discussion and improvement. On the other hand, this chapter argues that humanity needs to take serious actions now to build the infrastructure set out in the blueprint. This is because the threats to humanity already exist and are serious. One might revise a familiar phrase to communicate this point: *Think long-term, Act Now.*

Lastly, during the drafting of this book, the COVID-19 global pandemic bloomed. Rather than interweave commentary about the pandemic into each chapter, an Epilogue has been added at the end of the book. The piece captures thoughts about the pandemic and how it relates to various themes and topics addressed in the book.

Notes

1 The final draft of this book was prepared during the COVID-19 global pandemic of 2020. Please read the Epilogue for commentary on the pandemic as it relates to various themes and chapters of this book.

2 R. Lind, K. Arrow, G. Corey, P. Dasgupta, A. Sen, T. Stauffer, J. Stiglitz, and J. Stockfisch. 2011. *Discounting for Time and Risk in Energy Policy.* RFF Press, New York. https://doi.org/10.4324/9781315064048
3 C. Lindblom. 1959. The Science of "Muddling Through". *Public Administration Review,* 19, 2, 79–88. https://doi.org/10.2307/973677
4 Please see Chapter 8 for more on this observation.
5 B. E. Tonn, F. Conrad, and A. Hemrick. 2005. Cognitive Representations of the Future: Survey Results. *Futures,* 38, 810–829.
6 Brundtland Commission. 1987. *Our Common Future.* Oxford University Press, Oxford.
7 H. Daly. 1997. *Beyond Growth: The Economics of Sustainable Development.* Beacon Press, Boston, MA.
8 D. Meadows. 1994. *Envisioning a Sustainable World.* Third Biennial Meeting of the International Society for Ecological Economics, San Jose, Costa Rica, October 24–28.
9 P. Hawken. 2017. *Drawdown: The Most Comprehensive Plan Ever Proposed to Reverse Global Warming.* Penguin Books, New York.
10 R. Rosen. 1985. *Anticipatory Systems. Philosophical, Mathematical and Methodological Foundations.* Pergamon Press, Oxford, 341.
11 R. Poli. 2010. An Introduction to the Ontology of Anticipation. *Futures,* 42, 769–776.
12 University of Maryland, Center for Disaster Resilience, and Texas A&M University, Galveston Campus, Center for Texas Beaches and Shores. 2018. *The Growing Threat of Urban Flooding: A National Challenge.* College Park: A. James Clark School of Engineering. Retrieved from https://today.tamu.edu/wp-content/uploads/sites/4/2018/11/Urban-flooding-report-online.pdf
13 T. J. Gordon. 1994. Trend Impact Analysis. In J. C. Glen (Ed.), *Future Research Methodology.* American Council for the United Nations University, Washington, DC. Retrieved from www.foresight.pl/assets/downloads/publications/Gordon1994-Trendimpact.pdf
14 It should be noted that in 1932, H. G. Wells urged universities to appoint professors of foresight. https://foresightinternational.com.au/wp-content/uploads/2015/09/Wells_Wanted_Profs_of_Fsight_1932.pdf
15 H. Kahn. 1960. *On Thermonuclear War.* Princeton University Press, Princeton, NJ.
16 B. De Jouvenel. 1967. *The Art of Conjecture.* Basic Books, New York.
17 A. Toffler. 1970. *Future Shock.* Random House, New York.
18 J. Nesbitt. 1982. *Megatrends: Ten New Directions Transforming Our Lives.* Warner Books, New York.
19 Al. Gore. 2013. *The Future: Six Drivers of Global Change.* Random House, New York.
20 P. Schwartz. 1991. *The Art of the Long View: Planning for the Future in an Uncertain World.* Doubleday, New York.
21 W. Bell. 1997. *Foundations of Futures Studies: History, Purposes, and Knowledge,* vol. 1 & 2, Transaction Publishers, republished in 2017 by Routledge, New York.
22 R. Slaughter (Ed.). 1996. *New Thinking for a New Millennium.* Routledge, New York.
23 J. Dator. 1994. Dogs Don't Bark at Parked Cars. *Futures,* 26, 1, 898–894; K. van der Heijden. 1996. *Scenarios: The Art of Strategic Conversation.* John Wiley & Sons, Chichester; P. W. van Notten, J. Rotmans, M. B. van Asselt, and D. Rothman. 2003. An Updated Scenario Typology. *Futures,* 35, 5, 423–443; G. Wright and G. Cairns. 2011. *Scenario Thinking: Practical Approaches to the Future.* Palgrave Macmillan, New York; L. Borjeson, M. Hojer, K. Dreborg, T. Ekvall, and G. Finnveden. 2006. Scenario Types and Techniques: Towards a User's Guide. *Futures,* 38, 723–739; M. Amer, T. Daim, and A. Jetter. 2013. A Review of Scenario Planning. *Futures,* 46, 23–40.

24 In addition to the authors who published in *Futures*, the ideas presented in this book benefited from the extraordinary thoughts of numerous influential authors whose expertise span a multiple of disciplines. Here is a selection of the authors and their pieces that have shaped my thinking over the years: Peter Schwartz – *The Art of the Long View*; Ronald Heifetz – *Leadership Without Easy Answers*; Jared Diamond – *Guns, Germs and Steel*; Daniel Kahneman and Amos Tversky – *Prospect Theory: An Analysis of Decision Under Risk*; Herbert Simon – *Rational Decision Making in Business Organizations*; Joseph Campbell – *The Power of Myth*; Carl Jung – *The Concept of the Collective Unconscious*; Glenn Shafer – *A Mathematical Theory of Evidence*; Peter Senge – *The Fifth Discipline*; Charles Lindbloom – *Muddling Through*; Jane Jacobs – *The Death and Life of Great American Cities*; E. F. Schumacher – *Small Is Beautiful*; Paul Hawken – *Natural Capitalism*; Lewis Mumford – *The City in History*; Karen Armstrong – *Jerusalem*; Garrett Hardin – *Tragedy of the Commons*; Martin Luther King – *Letter from Birmingham Jail;* Robert Putnam – *Bowling Alone*; Robert Hellah – *Habits of the Heart*; Donella Meadows – *Thinking in Systems*; Ray Oldenburg – *Great Good Places*; Abraham Maslow – *Hierarchy of Needs*; Edward Tufte – *Envisioning Information*; Howard Gardner – *Multiple Intelligences*; Tom and David Kelly – *Creative Confidence*; William Julius Wilson – *When Work Disappears*; Edward O. Wilson – *Consilience: The Unity of Knowledge*; Martin Rees – *Our Final Hour*.

25 The field of futures studies does not offer a universally accepted definition of 'future generations'. I tend to interpret the concept in terms of children: my children's children, etc., etc., etc.

26 R. Slaughter. 1993. The Knowledge Base of Futures Studies. *Futures*, 25, 227–233. https://doi.org/10.1016/0016-3287(93)90134-F.

27 Mary Wollstonecraft Shelly published the book, *The Last Man*, in 1826, that could be considered a harbinger of worries about human extinction. See Chapter 4.

28 U.K. Treasury. 2006. *Stern Review on the Economics of Climate Change*. UK Treasury, London.

29 N. Bostrom. 2002. Existential Risks: Analyzing Human Extinction Scenarios and Related Hazards. *Journal of Evolution and Technology*, 9, 1, 1–31.

30 United Nations. 17 Sustainable Development Goals. Retrieved from www.un.org/sustainabledevelopment/sustainable-development-goals/

31 R. Kurzweil. 2005. *The Singularity Is Near: When Humans Transcend Biology*. Penguin Books, New York.

2 A serendipitous trip through future time

Like most people, when I hear the word *future*, or when I think about *futures*, my thoughts are often confined to my personal sphere. I focus on relationships, health, finances, etc., not only of my own, but those of my spouse, children, and other relatives and close friends and colleagues. I enjoy anticipating the release of a new movie or attending a concert. I pine for a future where my vote will make a difference. I do pretty well at anticipating events and planning my time at least a few days in advance.[1]

However, as one who has contributed to the futures literature for a number of years, when I hear the word *future* my thoughts also veer off in many other directions, where the time horizons of my thoughts could be decades, centuries, millennia, or even longer. Hearing the word prompts a profound reframing of my perceptions of and relationships with physical and social realities. This reframing is in some ways as pronounced and jarring as the new reality experienced by Neo, the protagonist in the movie *The Matrix*. Upon taking the red pill, Neo came to realize that the world he thought he lived in was simply a computer-generated simulacrum. In this world, his body – and those of hundreds of millions of others – was used by Earth's AI [artificial intelligence] overlords to power the world's computer infrastructure. After taking the pill, he was forced to see the world in a more fundamental way and to deal with reality in a more fundamental manner. Although it was a struggle, he eventually found the courage to accept and deal with the actual world. One can consider that the 'red pill' is an apt metaphor for this book because after reading this book, every aspect of reality, perceived through futures-eyes, could look decidedly different.

My goal for this chapter is to similarly set the stage for mindful transformation, from thinking about futures in the present tense and personal sense to thinking about futures in an expansive, biggest picture possible way. I should note that I am using the plural *futures* instead of the singular *future*, as already touched upon in Chapter 1.

Figure 2.1 presents a graphical representation of much of what I think of when, as a futurist, I hear the word 'future'. The timeline extends past one billion years. This span of time includes events we can anticipate and also achievements by humanity that can be considered absolutely necessary for

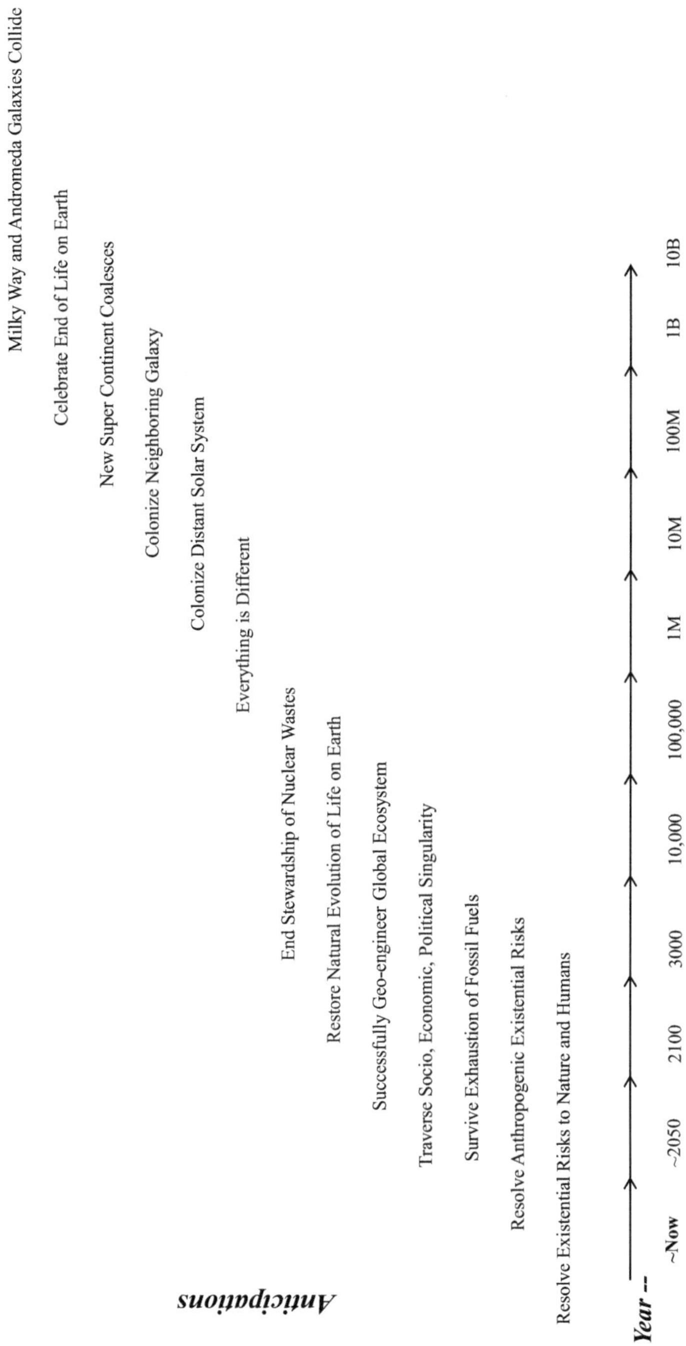

Figure 2.1 Anticipations and Aspirations Through Future Time

our survival and also gloriously aspirational. Let's eschew temporal linearity at this point in favor of storytelling and begin this exploration toward the end of the timeline, one billion years into the future, then work our way back to 50 years into the future before once again greatly expanding our time horizons. Consider this an exercise in time travel![2] If you feel you need a bit more preparation before heading on this journey through futures, please try the exercise contained in Exhibit 2.1 at the end of this chapter before embarking on this journey.

Year One Billion – One billion years from now, give or take 500 million years, the earth will become uninhabitable to life of any sort.[3] This is because, paradoxically, as the sun nears the end of its life, scientists theorize that it will expand tremendously, eventually engulfing the innermost planets in the solar system, including the earth. Long before this actually happens, intense heat from the sun's expansion will boil off the oceans and scorch the planet. Life on earth will not survive this inferno. From a futures perspective, this is a hard stop for life on earth, though not necessarily for earth life.

I understand that one billion years is a long time and extraordinarily challenging to grasp when we are consumed with meeting the demands of everyday life! However, in the grand scheme of things, one billion years is not that long. For example, one billion years is only a fraction of the current age of the universe, which is 13.8 billion years,[4] and only about a quarter of the age of the earth itself, which is about 4.5 billion years.[5] One billion years is even short compared to how long ago life first sprang up on the earth, about 3.8 billion years ago (estimated).[6] From this perspective, resorting to an American football analogy, life on earth is actually in its fourth and final quarter, even though to humans it seems as though we are still early in the first minute of the first quarter of our existence in this universe! In any case, the Year One Billion provides the penultimate planning horizon. I believe that we should set goals of maintaining life on earth until the earth becomes uninhabitable while also finding other homes in the universe to transcend the oblivion of the earth.[7]

Take-Home Point #1: The earth will become uninhabitable in one billion years.

Year Ten Million – Even though one billion years is, well, one billion years into the future, humanity should already be thinking ahead to that bitter end and making plans to colonize other solar systems in our galaxy. We should be looking for solar systems with suns that are expected to have lifetimes well beyond the lifetime of our sun. We are making good progress in detecting planets that are orbiting distant suns.[8] So far, most do not appear to reside in the Goldilocks zone that the earth inhabits around our sun: not too close as to be too hot and not too far as to be too cold. Also, most appear to be gaseous giants like Jupiter instead of planets the size and composition of the earth that could more easily support life. Still, the search for inhabitable planets is really still in the initial stages. Seemingly potentially inhabitable homes have been found. This is the easy part.

The hard parts are getting there and then terraforming planets to meet our needs. Travel time alone could take tens of thousands of years. Let's do a simple thought experiment. Let's assume that the fastest space-faring vehicle that humanity could produce moves at 0.1% of the speed of light or about 670,000 miles per hour. If a likely new home for earth life is 1000 light years away, it will take about one million years to reach this new planet. It would take 1000 years for information to return to the earth, from an extraordinarily durable probe, for example, to inform humanity that yes, we should endeavor to terraform this new habitat. Another million years could pass before another contingent from the earth makes its way to this planet and maybe another million years for terraforming and then another million years for actual intelligent life to move to this new planet and one can quickly see that year ten million is not an unreasonable date for purposes of discussion for the placement of this event on our timeline. Of course, this date would be then also be the minimum planning target for sustaining life on the planet earth.

This achievement could add a few additional billions of years to humanity's existence if these planets are located in solar systems whose suns have longer lifetimes. Still, this achievement cannot save all life on earth. It is unlikely that the resources will be available to move all humans off the planet earth, much less all animal, plant, and microbiotic life. Thus, we will still be celebrating the end of life on earth about one billion years from now, even if the earth becomes unable to sustain humanity.

> *Take-Home Point #2: Establishing homes elsewhere in the galaxy could take millions of years.*

Year 2500 – Let us take another perspective on ten million years – an energy perspective. From an anthropocentric standpoint, concentrated forms of usable energy are needed to power an advanced civilization of any form. Energy powered the Industrial Revolution and the Information Technology Revolution and will power all succeeding science, technology, and economic revolutions. Consumption of energy infuses our built architecture, transportation systems, manufacturing plants, and telecommunication systems. The preponderance of our energy still comes from non-renewable fossil fuels.[9] Given modern capitalist system modes of ownership, owners of fossil fuel resources will likely continue to thwart limits to fossil fuel consumption, limits that are desperately needed to reduce greenhouse gas emissions and other emissions that pollute our air and water. They believe, as owners of these resources, that they deserve maximum value for their resources, and, of course, they will not get maximum value until these resources they own are extracted and consumed. A similar argument can be made about the owners of nuclear resources and nuclear power plants and technologies.

Potential impacts of catastrophic climate change from the consumption of all the world's fossil fuels are discussed later, at year 2100. For the moment, let us first focus on this non-renewable resource. How many more years can fossil fuels be consumed at about the current rate until, for all intents and purposes, this resource is fully depleted, catastrophic climate change notwithstanding? A review of various estimates related to coal, natural gas, and oil reserves puts the period at about 500 years – around the Year 2500, with coal being the last fossil fuel standing.[10] Oil will be the first to run out. Many believe that the world has hit or will soon hit its peak oil production (for all our remaining history on this planet).[11] It is certainly the case that discoveries of new oil deposits has severely declined in number and in barrels of potentially recoverable oil since the 1960s.[12]

Despite the easily anticipatable decline of big oil, many energy analysts are content that natural gas supplies seem to be plentiful for the next decade or two, that we may not move past peak oil production until 2050 or so, and that, yes, we have plentiful supplies of coal for the next several hundred years. But, now, let us put on our futures hats. Five hundred years is but a drop in the bucket of time compared to a sustainability planning horizon of at least ten million years. Contentment is not an appropriate emotion from our new futures perspective.

Additionally, in many ways, 500 years also seems to be a short time to completely transition away from fossil fuels. It seems a short time to build storage for nuclear waste, to ramp up other renewable sources, and to devise the smart grid capacities to handle the world's increasing energy demands. It seems a short time to completely wean the transportation sector off of fossil fuels, to electrify the transportation infrastructure. It is a short time to replace all of our buildings, to learn how not to build roads and buildings out of concrete, whose processes emit prodigious amounts of greenhouse gases (GHG) into the atmosphere each year. Being sanguine that the world has ample supplies of fossil fuels and can quickly transition to other sources of energy and infrastructures is embarrassingly shortsighted given the timescales addressed in this book.

Take-Home Point #3: Depletion of non-renewable resources over the next couple of centuries could impact human civilizations for thousands and millions of years into the future if the difficult transition to a portfolio of non-fossil fuels is not started now and completed expeditiously and effectively.

Year 2100 – The earth appears to be on an irreversible path toward further climate change over the next century. We are already witnessing: increases in mean global and regional temperatures; sea-level rise; disappearances of glaciers; changes in geographic locations of the habitats of flora and fauna; and increases in the severity of extreme weather events, from droughts to storms.[13] It can be argued that climate change is an existential risk, meaning that it could lead to the extinction of the human race. It can also be argued that climate change is just one of many existential risks facing humanity. The next chapter, Chapter 3, takes an in-depth look at a comprehensive set of such risks. The important points here are that existential risks like climate change ought not be ignored and that those seemingly being caused by humanity need to be dealt with much sooner than later. One can anticipate that this sanguinity could result in globally catastrophic collapse of our technological infrastructure, our economies, and even in the world's population.[14]

> *Take-Home Point #4: Reducing existential risks is an essential component of long-term sustainability thinking. We can anticipate our world being in the throes of catastrophic climate change, maybe even at an unacceptably high risk of human extinction, if by the year 2100 we have not completely solved our GHG emissions problem.*

Year 2050 – In addition to climate change, there are other existential risks that need to be addressed sooner rather than later. These risks also have their roots in technology but are not so overtly destructive in nature. These risks are related to the integrity of life on earth (i.e., what nature is and means) and what it means to be human. These risks stem from the development and use of technologies that alter the genetic composition of species.

Genetic engineering technologies have produced numerous benefits for humanity. Crops of various sorts are more productive and resilient to pests, disease, and fluctuations in weather.[15] A new technology, called clustered regularly interspaced short palindromic repeats (CRISPR, pronounced *crisper*) has emerged recently that greatly enhances the ability of scientists to manipulate the genetics of species.[16] Specifically, this technology allows scientists to insert genetic sequences, either natural or synthetically developed, into the genomes of species in ways that ensure that these sequences are passed down to succeeding generations. This technology is often referred to as *gene drive technology*. Researchers are currently exploring driving genes into Asian carp and brown rats to eliminate their invasion of non-native habitats and into mosquitoes to reduce malaria.[17] The technology is also being considered to drive genes into American chestnut trees to protect them from the chestnut blight.[18]

The existential issues are these: at what point does human manipulation of other species' genomes pass a threshold to where the naturalness of nature ceases to exist, to where life on earth loses its dignity and essentialness, to where life on earth is a means only for the achievement of humanity's goals and not its own entity? With respect to long-term futures, what is it then that we want to preserve for future generations, if anything? Will natural evolution be so constrained by human development and rendered impotent by genetic interventions during the Anthropocene that for all intents and purposes it will cease to exist and if so, is this an acceptable potential future for earth life? We need to answer these questions much sooner than later because a raft of changes to the earth's species are close to being unleashed around the globe, many of which will be irreversible.

These same questions are also pertinent with respect to the future of humanity. CRISPR technology could have much impact on the human species in the next decade or two. For example, clinical trials have already begun to use CRISPR to fight cancer and blood disorders.[19] One can imagine that other genetic engineering technologies, combined with nano-, information, and cognitive technologies, could be used to ultimately create transhumans or even new species of humanoids. My thinking about long-term futures of humanity is based on an assumption that humanity retains its *essential humanness* over time. At this point in time, I cannot completely describe what essential humanness means, but it probably excludes potential futures where 'humans' exist only as computer constructs. I cannot say whether essential humanness is preserved in cases where humans live for centuries aided by technologies that will be able to grow new organs for transplant, undergo drug and other treatments to stave off dementia, enhance cognitive abilities and memory, and are created from genetically manipulated eggs, sperm, and fetuses that virtually eliminate inheritable genetic diseases and build protections against common environmental risks (e.g., to manipulate skin cells to withstand cancer-causing impacts of the sun)? I do believe, however, that these capabilities will be available to humans in the near future. I also believe that such changes to humans are also likely to be irreversible, meaning that lifeforms that descend from you and me will not be able to voluntarily revert to our most humble beingness. Thus, we need to practice responsible foresight and deal with these deep existential questions about nature and humanness within the next several decades.

> *Take-Home Point #5: Over the course of several decades, humanity must be wise about the development and use of technologies that could pose fundamental existential risks to nature and itself.*

Year 3000 – We have discussed how advanced technologies could threaten the essential character of nature and humans and the many anthropogenic-based threats that could destroy humanity and life on earth. We also addressed energy issues, which need to be resolved within the next several centuries. One can imagine that to address these and other challenges, humanity will have to evolve new economic and political systems, new lifestyles, new cultures and forms of identity, and maybe even new religions and belief systems. By then, we should have achieved true long-term environmental sustainability, a world where environmental regulations may no longer be necessary. If all of these changes are achieved, the world will be vastly different from what we experience today. Everything will be different.

By Year 3000, humanity should work toward, plan for, and successfully pass through the proverbial socio-economic *Singularity* (to borrow terminology from Ray Kurzweil[20]). Chapter 9 describes in more depth the socio-cultural transformations I have in mind.

Take-Home Point #6: Around the year 3000, one of two things will have happened: civilization will have collapsed, or every aspect of society will have radically changed to achieve long-term sustainability.

Year 4000 – Serious discussions are taking place in scientific circles around the world on the topic of geoengineering. The idea is to put in place technologies to ameliorate global warming and thereby prevent the most calamitous consequences of global climate change. Ideas range from seeding the oceans with iron to promoting the accelerated growth of carbon-consuming microorganisms to spreading sun-reflecting sulfur particles in the atmosphere to placing giant sun reflectors in space.[21] Many are quite worried, rightly I believe, about the potentially negative and even more destructive unintended consequences of these schemes.

Implementing risky geoengineering schemes simply because humanity does not have the wherewithal to change our behavior in the near term to reduce consumption of fossil fuels seems misguided at best. Again, it also seems to mean that we intend to rely only on risky technological solutions instead of working toward and through the Year 3000 socio-cultural Singularity, where humanity gladly changes behaviors and institutions to ensure our journey into the future and does not rely on the solution of least political resistance that may inadvertently entail the most overall risk to humanity.

On the other hand, eventually, humanity will need to learn to manage the earth's climate. We will not survive into the distant future if global warming turns the earth into an uninhabitable hothouse like Venus.[22] Likewise, humanity also will not flourish and may not survive a period of extreme global cooling, especially a snowball earth scenario where ice and glaciers from the north and south could eventually come close to meeting at the equator.[23] It is proposed that humanity takes it time to learn how to geo-engineer the earth's climate so as to maximize the benefits to all species and generations while avoiding potential unintended existential risks. Year 4000 seems like a reasonable time period to achieve this goal. This also means that geoengineering is off the table as a solution to our current global climate change problems.

Take-Home Point #7: Eventually, the earth will need to be geo-engineered to ensure long-term sustainability of life on earth.

Year 5000 – The goal for Year 5000 is to thoroughly restore natural evolution to the planet earth. In other words, the era of the Anthropocene will have ended to be replaced by a new era. Humans have appropriated, changed, and manipulated, some might even argue geo-engineered, nature for its own benefit for many millennia. It can be argued that few if any terrestrial ecosystems exist that can be considered natural (i.e., unchanged and unfettered since before the emergence of Homo sapiens ~125,000 years ago). Over the past 20,000 years, humans have been implicated in the extinction of scores of megafaunas[24] and in recent decades as the perpetrators of a sixth massive extinction.[25] Additionally, human interventions in nature through genetic engineering further constrain the process of natural evolution that resulted in an explosion of wondrous species diversity in previous epochs.[26]

My vision for the earth in the Year 5000 is a world where large swaths of the earth and oceans have reverted to a natural state. This state would not resemble nature before humans evolved, because that world is irretrievably lost. This state is probably better described as a resetting of nature to reboot natural evolutionary processes. Humans would have vastly reduced their agricultural footprints across the globe and even intricately designed nature to co-evolve with humans' highly sustainable and environmentally friendly settlements. These plans would synthesize with the results of the geoengineering efforts achieved in the previous millennium. Humans would husband and care for the balanced natural processes until the Year One Billion. Over this time, at least to the time of colonization of space, we could learn more about life and its ability to adapt and be resilient, which could then aid terraforming exercises. I believe that this is both an ethical goal, to stop interfering with natural processes, and a practical goal because allowing natural evolution reduces unintended consequences of our own manipulation of nature.

Take-Home Point #8: It is important to restore natural and unfettered evolution on the planet earth.

Year 10000 – The Year 10000 has special significance for humanity. It is around this year when most of the nuclear wastes generated by the world's nuclear power plants in the twentieth, twenty-first, and probably twenty-second centuries will have become harmless to humans and other life on earth. It is interesting that the field of futures studies, at least the branch that addresses concerns about obligations to future generations, arose, in part, because of the prospect of having to deal with the specter of nuclear wastes for thousands of years. It seemed to many that it was unfair for current generations to bequeath to future generations the risks and responsibilities for managing these radioactive residues.[27]

It is also interesting to note that some individuals thought inevitable upheavals or displacements of human settlements could mean that the location of nuclear waste repositories would be lost to human memory within 10,000 years. Because of this potentiality, warning signs have been embedded in indestructible signs to warn future generations.[28] Of course, if humanity through the course of time loses track and memory of these repositories, humanity will not have survived and met the milestones discussed immediately earlier. In other words, if the nuclear waste repositories become dust-covered ruins to be discovered by future generations, it is probable that there will be no future generations in existence to discover the said ruins. This scenario certainly runs counter to the optimistic and empowering futures envisioned by this book! Therefore, as addressed in Chapter 9, it is important to have stable, perpetual institutions to manage these wastes over ten millennia.[29]

Take-Home Point #9: Humanity needs to successfully manage nuclear waste repositories for ten millennia or more.

Year One Million – A group of writers coalesced to create a book about what they imagined life would be like in the Year One Million.[30] All of the contributors stated in one way or another that life would be exceedingly different at this milestone in humanity's history. Among many vivid images are these:

- Humanity will be in regular communication with life elsewhere in the universe;
- Human lifespans will approach immortality;
- Jupiter will be mined for deuterium and humanity will dismantle asteroids for minerals;
- Our understanding of physics and reality will progress to where we will understand superluminal travel;
- A Matrioshka brain will be built;[31] and
- Everything will change but numbers and laughter.

Year Fifty Million – Assuming humanity successfully colonizes the Milky Way Galaxy, it is natural to assume that humanity will explore other galaxies for new homes. Of course, this will take even more time, even with exceedingly advanced technologies. For the sake of this story, let us assume that humanity plans on colonizing other galaxies by the Year Fifty Million.

Take-Home Point #10: It should be a goal of humanity to find homes in neighboring galaxies.

Year One Hundred Million – While humans are actively exploring space, geological processes on the earth will have created discernable changes in the earth's geography. For example, by this point in time, some believe that a new super continent, which some are calling Amasia, will coalesce in the northern polar region.[32] Whether the existing continents collide there or someplace else on the earth, they will continue to move and at these timescales, their movement will impact every aspect of life on earth, from the global climate to the inhabitability of various land masses to the inevitable mash-up of geographies we now refer to as countries.

Beyond Year One Billion – What happens beyond the Year One Billion? Humanity will have colonized other solar systems in our own galaxy and also in neighboring galaxies. Hopefully, our distant descendants will have considered the prediction that the Milky Way and Andromeda Galaxies collide around four billion years from now. And then what? Will humanity eventually build Type III Kardashev civilizations, where humanity is able to control all energy sources in an entire galaxy?[33] Will humanity be able to colonize other universes to avoid the loss of functional energy sources in our own universe? Or even create other universes? Maybe, maybe not. I really have no idea. But I would love to see what happens one billion plus years from now! Since that seems impossible, it is my fervent hope that our journey continues to this time and beyond!

Take-Home Point #12: Humanity should set its sights on grand achievements beyond the death of our sun.

Observations – I hope that now when the reader hears the words future or futures, the experience provokes expansive thoughts. There are important global milestones within our personal lifetimes, such as addressing oblique insidious threats to nature and humanity to reducing more overt threats to humanity's existence from climate change and transitioning the world's energy systems away from fossil fuels. Plans and activities are needed to successfully transverse the Year 3000 socio-cultural singularity and meet the Year 4000 and Year 10000 goals of successfully geoengineering the earth's climate and managing nuclear wastes. There is no better time than the present to begin to meet goals for resetting 'natural' evolution to the planet and finding new homes for earth life amongst the stars.

Figure 2.1 displays only a small fraction of other major events and challenges that one can anticipate through careful and systemic reflection. For example, one might be tempted to add various scientific and technological breakthroughs to the list. Developing a testable 'theory of everything,', explaining dark energy and matter, and discovering life beyond earth come quickly to mind. Quantum computing, self-replicating nanotechnologies, fusion energy, and super-artificial intelligences are also possible to be in our future. With respect to culture and society, one may also anticipate the natural waning of conventionally understood races, ethnicities, and languages as populations continue to mix (voluntarily and involuntarily, because of climate change migration, for example) to be replaced with fresh new notions of self-identity and culture. Over the course of time, our understanding of science and the humanities will find further consilience to deepen our understanding of the meaning and purpose of life.[34]

One may also anticipate a host of additional environmental problems in addition to climate change. For example, soil degradation is already recognized as a major problem affecting our agricultural base.[35] Imagine the scale of this problem over the course of another thousand years if current trends continue. The world's industries annually generate millions of tons of air and water pollutants, which are deposited in our waters, soils, foods, and ultimately in our bodies and other species of life. What would the earth's soils and waters and biota look like after 1000 years of such chemical depositions? One can also anticipate that over this time span all nine of the planetary boundaries proposed by a group of researchers led by the Stockholm Resilience Centre could be dangerously crossed.[36]

As a counterpoint, this chapter demonstrates that effective and useful futures-thinking does not have to involve down-to-the-year technological forecasts and other predictions. Futures studies are often dismissed or outright derided because so-called futurists have been so errant with respect to their forecasts regarding technology. For example, the chairman of IBM once admitted to underestimating demand for new IBM computers.[37] Others have oversold the benefits of nuclear power and completely missed on the internet and even iPhones.[38] Energy prices and the stock market are notoriously hard to predict but people still do, which I refer to as myopic futuring, which then, in turn, soils the environment for those who focus on the larger perspectives captured by this book.

This is not to say that technologies are not important now and will not shape the future. Of course, they will. The point is that one need not have to predict the date when a very specific new technology will emerge in order to take a journey through the future set out earlier. For example, around 2050 the nature of nature and the nature of being human will both be threatened, if that is not already the case. Yes, CRISPR technology will be complicit, but other technologies may also emerge to amplify CRISPR's potential impacts on humans and nature writ large. We can see and feel these technology trends; we do not necessarily have to forecast specific new technologies to anticipate these fundamental risks.

Conversely, new and as yet unimaginable technologies may wreak havoc with the timeline presented in Figure 2.1 but not its essential components. For example, new technologies could be developed to render stored nuclear wastes harmless long before their emissions naturally decay into harmlessness,[39] but that does not mean that this issue should not be part of the timeline. Humanity may actually discover the means for faster-than-light (FTL) travel, thereby significantly moving up timelines for finding other homes in our and neighboring galaxies, but this does not mean that we ought to drop these items from our anticipations with respect to our journey through space and time.

Underlying the arguments presented earlier is an essential but implicit assumption: that we should care about future generations of humans and the future of life on earth. It assumes that the journey of life through time not only is worth living but also preserving for all of time. For many people, the question of why we should care for future generations is an open question; this question is addressed in Chapter 5. Our perpetual obligations for ensuring the journey are addressed in Chapter 6. First, Chapters 3 and 4 take on the issue of human extinction and our agency to prevent our demise, respectively.

As a last point, the purpose of this chapter is to encourage futures-oriented thinking within the biggest picture possible. This intent is to expand our horizons without reducing concerns about the near term. There are many things to worry about in the coming years, including pandemics, invasions of autonomous vehicles and drones, immediate impacts of climate change (e.g., urban flooding), rising suicide rates, the opioid epidemic, and the perplexing reemergence of invective intolerance, nationalism, and racism. A maturing human civilization needs to learn to simultaneously take on both the near-term and longer-term challenges.

Let's end this chapter with some positive thoughts from Peter Ward:

> *My own view is that we will successfully negotiate the hazards threatening our species. We will not kill ourselves off. We will not die off from disease. We will wax and wane with all manner of climate change, asteroid impacts, runaway technology and evil robots. We will persevere . . . Perhaps this view that we are unkillable – at least as a species – is naïve. But even if we are to live as long as an average mammalian species – between 1 and 3 million years – we still have huge stretches of time left, for our species is barely a quarter of a million years old. And who says we are average? My bet is that we will stick around until the very end of planetary habitability for this already old earth.*[40]

Exhibit 2.1 A futures-thinking exercise

The purpose of this exercise is to help you become more comfortable in thinking about futures. The approach followed here is to first have participants immerse themselves in past times and settings, and then take them into the future.[41] This exercise has two immersive in the past steps and then a scenario-writing task.

First, imagine yourself living in a hunter-gatherer society many thousands of years ago. It is a summer day. What do you see around you? Who are you living with and how is their health? What possessions do you have? What do you smell? What are you wearing? How does your group think about the future? Take about five minutes to ponder these questions.

Second, imagine yourself living in a tenement building in a major city in the mid-1800s, such as New York City or London. You happen to be in your unit in your building in the early evening on a winter day. What do you see around you? Who are you living with and how is their health? What possessions do you have? What do you smell? What are you wearing? How does your group think about the future? Take about five minutes to ponder these questions.

Third, imagine a household living in an apartment building in an urban area 30 years from now. Think about the following questions:

- Where is this apartment located (i.e., city, country)?
- How many individuals are in the household? What are their ages?
- Do the individuals have health issues? Are they disabled?
- Are they working or in school? Or unemployed?
- What is their income? Are they poor?
- How do they self-identify with respect to gender, race, and ethnicity?
- What do you see, smell, hear?
- What are your possessions? What is the condition of your apartment?

Then create a scenario that describes the lives of the people living in the apartment. Following Schwartz, a scenario is a story that depicts the most important aspects of life in the future that is based upon the most important trends you believe will impact those lives.[42] The story could be fashioned as a diary entry or a news story, for instance. Each scenario should have an evocative name.

For this exercise, give yourself five minutes to work out a description of the household and 15 minutes to create the scenario. All you need is pen and paper or perhaps a whiteboard and a marker. It is also suggested that you find a partner for this exercise. When you are finished, take a few minutes to reflect upon this exercise. Did the exercise relieve any anxieties you may have felt thinking about the future? What did you learn from studying futures through this exercise?

Notes

1 In fact, planning is an essential aspect of human behavior. See G. A. Miller, E. Galanter, and K. Pribram. 1960. *Plans and the Structure of Behavior.* Henry Holt and Co., New York.

2 Since I just mentioned time travel, let me say that in my opinion, we are living in a second golden age of science fiction. Sci-fi authors who have influenced my thinking and work in futures studies include: Ian Banks, William Gibson, Orson Scott Card, Robert Sawyer, Neil Stephenson, Charles Stross, John Scalzi, Connie Willis, and Neil Asher. As a young adult, I do have to thank Isaac Asimov, Frank Herbert, Phillip K. Dick, and Robert Heinlein for their wonderful books. Lastly,

I would also like to add *The Last Man* by Mary Wollstonecraft Shelley and *Islandia* by Austin Tappan Wright to this list of fiction that I treasure.

3 P. Ward and D. Brownlee. 2002. *The Life and Death of Planet Earth.* Henry Holt, New York.

4 National Aeronautics and Space Administration. 2012. *How Old Is the Universe?* Retrieved from https://map.gsfc.nasa.gov/universe/uni_age.html.

5 United States Geological Survey. 2016. *The Age of the Earth.* Retrieved from https://geomaps.wr.usgs.gov/parks/gtime/ageofearth.html.

6 E. Nisbet and N. Sleep. 2001. The Habitat and Nature of Early Life. *Nature*, 409, 1083–1091.

7 B. E. Tonn. 1999. Transcending Oblivion. *Futures*, 31, 351–359.

8 Planetary Habitability Laboratory @ UPR Arecibo. 2020. Retrieved from http://phl.upr.edu/projects/habitable-exoplanets-catalog

9 One could maybe argue that, given the time frames considered herein, fossil fuels could indeed be replenished by natural means, maybe in the 300-million-year time frame. Unfortunately, this does not help humanity get to the year 3000 without collapse or provide the energy infrastructure for extraterrestrial colonization.

10 T. Covert, M. Greenstone, and C. Knittel. 2016. Will We Ever Stop Using Fossil Fuels? *Journal of Economic Perspectives*, 30, 1, 117–138.

11 D. Helm. 2011. Peak Oil and Energy Policy – A Critique, *Oxford Review of Economic Policy*, 27, 1, 68–91.

12 R. Miller and S. Sorrell. 2014. The Future of Oil Supply. *Philosophical Transactions of the Royal Society A*, 372. http://dx.doi.org/10.1098/rsta.2013.0179

13 Intergovernmental Panel on Climate Change. 2018. *Global Warming of 1.5°C.* An IPCC Special Report on the Impacts of Global Warming of 1.5°C Above Pre-industrial Levels and Related Global Greenhouse Gas Emission Pathways, in the Context of Strengthening the Global Response to the Threat of Climate Change, Sustainable Development, and Efforts to Eradicate Eoverty. V. Masson-Delmotte, P. Zhai, H.-O. Pörtner, D. Roberts, J. Skea, P. R. Shukla, A. Pirani, W. Moufouma-Okia, C. Péan, R. Pidcock, S. Connors, J. B. R. Matthews, Y. Chen, X. Zhou, M. I. Gomis, E. Lonnoy, T. Maycock, M. Tignor, and T. Waterfield (Eds.). In Press. Retrieved from www.ipcc.ch/site/assets/uploads/sites/2/2019/06/SR15_Full_Report_Low_Res.pdf

14 J. Diamond. 2005. *Guns, Germs, and Steel: The Fates of Human Societies.* W.W. Norton, New York.

15 National Academies of Sciences, Engineering, and Medicine; Division on Earth and Life Studies; Board on Agriculture and Natural Resources; Committee on Genetically Engineered Crops: Past Experience and Future Prospects. 2016. Genetically Engineered Crops: Experiences and Prospects. National Academies Press (US), Washington, DC.

16 J. Doudna and S. Sternberg. 2017. *A Crack in Creation: Gene Editing and the Unthinkable Power to Control Evolution.* Mariner Books, New York.

17 T. Harvey-Samuel, T. Ant, and L. Alphey. 2017. Towards the Genetic Control of Invasive Species. *Biological Invasions*, 19, 1683–1703.

18 A. Newhouse, A. Oakes, H. Pilkey, H. Roden, T. Horton, and W. Powell. 2018. Transgenic American Chestnuts Do Not Inhibit Germination of Native Seeds or Colonization of Mycorrhizal Fungi. *Frontiers of Plant Science*, 9, 1046.

19 T. Saey. 2019. Science News, August 14. Retrieved from www.sciencenews.org/article/crispr-gene-editor-first-human-clinical-trials

20 R. Kurzweil. 2005. *The Singularity Is Near: When Humans Transcend Biology.* Viking, New York.

21 E. Kintisch. 2010. *Hack the Planet: Science's Best Hope – or Worst Nightmare – For Averting Climate Catastrophe.* John Wiley & Sons, Hoboken, NJ.

22 M. Way, A. Del Genio, N. Kiang, L. Sohl, D. Grinspoon, I. Aleinov, M. Kelley, and T. Clune. 2016. Was Venus the First Habitable World of Our Solar System? *Geophysical Research Letters*, 43, 8376–8383. https://doi.org/10.1002/2016GL069790.

23 P. Hoffmann and D. Schrag. 2000. Snowball Earth. *Scientific American*, 282, 1, 68–75.
24 J. Diamond. 2005, ibid.
25 E. Kolbert. 2014. *The Sixth Extinction: An Unnatural History*. Henry Holt and Company, New York.
26 S. J. Gould. 1989. *Wonderful Life: The Burgess Shale and the Nature of History*. W.W. Norton, New York.
27 D. MacLean, D. Bodde, and T. Cochran. 1981. *Introduction to Conflicting Views on a Neutrality Criterion for Radioactive Waste Management*. College Park: University of Maryland, Center for Philosophy and Public Policy. K. Shrader-Frechette. 1994. Equity and Nuclear Waste Disposal. *Journal of Agricultural Environmental Ethics*, 7, 133–156.
28 G. Benford. 2000. *Deep Time: How Humanity Communicates Across Millennia*. Avon Books, Harper Collins, New York.
29 B. E. Tonn. 2001. Institutional Designs for Long-term Stewardship of Nuclear and Hazardous Waste Sites. *Technological Forecasting and Social Change*, 68, 255–273.
30 D. Broderick (Ed.). 2008. *Year Million: Science at the Far Edge of Knowledge*. Atlas Publishers, New York.
31 This is a computer the size of a planet, essentially https://en.wikipedia.org/wiki/Matrioshka_brain
32 R. Harris. 2012. All Things Considered: 'Amasia': The Next Supercontinent? February 8. Retrieved from www.npr.org/2012/02/08/146572456/amasia-the-next-supercontinent
33 J. Creigton. 2014. The Kardashev Scale-Type I, II, III, &V Civilization, July 19. Retrieved from https://futurism.com/false-coloring; Kardashev Scale. 2020. *Wikipedia*, Wikimedia Foundation, March 2. Retrieved from en.wikipedia.org/wiki/Kardashev_scale.
34 E. O. Wilson. 1998. *Consilience: The Unity of Knowledge*. Knopf, New York.
35 J. Quinton, G. Govers, K. Van Oost, *et al*. 2010. The Impact of Agricultural Soil Erosion on Biogeochemical Cycling. *Nature Geoscience*, 3, 311–314. https://doi.org/10.1038/ngeo838
36 Stockholm Resilience Centre. 2020. The Nine Planetary Boundaries. *Planetary Boundaries*. Retrieved from www.stockholmresilience.org/research/planetary-boundaries/planetary-boundaries/about-the-research/the-nine-planetary-boundaries.html
37 Geek History. 2020. Urban Legend: I Think There Is a World Market for Maybe Five Computers. Retrieved from https://geekhistory.com/content/urban-legend-i-think-there-world-market-maybe-five-computers
38 J. Szcerba. 2015. 15 Worst Tech Predictions of All Time. *Forbes Billionaires*. Retrieved from www.forbes.com/sites/robertszczerba/2015/01/05/15-worst-tech-predictions-of-all-time/#4d1add1a1299
39 M. Wald. 2013. A New Vision for Nuclear Waste. *Technology Review*, December 1. Retrieved from www.technologyreview.com/s/403437/a-new-vision-for-nuclear-waste/
40 P. Ward. 2001. *Future Evolution: An Illuminated History of Life to Come*. Henry Holt, New York, 167.
41 O. Markley. 1994. Experiencing the Needs of Future Generations. In Milton Moskowitz and Trans Pacific Bridge, Inc. (Eds.), *Thinking about Future Generations*. Institute for the Integrated Study of Future Generations, Kyoto, Japan, 215–231.
42 P. Schwartz. 1996. *The Art of the Long View: Planning for the Future in an Uncertain World*. Doubleday, New York.

3 Existential risks

Humanity faces a surfeit of problems. Many were touched upon in the previous chapter, such as climate change, storage of nuclear wastes, and availability of energy supplies. Ethnic violence, human trafficking, aging and shrinking national populations, lack of safe and affordable housing, and lack of sufficient potable water are current problems that also demand attention and resolution. Solving these problems in the near term will certainly benefit our descendants in the longer term. So will tackling problems that threaten the potential for our future generations to even exist. It is a category of risks that must be addressed if we are to protect humanity's journey through time and space within the biggest picture possible framework of this book. It is a category that I believe has not received its due attention from today's generations and policy makers. We need to summon the courage to face these risks, following Carl Sagan who stated: "*For me, it is far better to grasp the universe as it really is than to persist in delusion, however satisfying and reassuring.*"[1]

The chapter has two main sections. The first section addresses this question: is human extinction inevitable? Two collections of evidence seem to suggest that the answer to this question is yes. The second section addresses this question: what existential risks does humanity face? The answer to this question is quite a few. My own view is that humanity ought not submit to the lure of fatalism but to use the evidence in the first section as motivation to avoid this prophecy; and humanity should energetically and steadfastly take on even the most daunting challenges to its existence.

Is human extinction inevitable?

Two main arguments have been made that human extinction is inevitable. The first is that the vast majority of species that have ever existed on the earth have become extinct. There have been five mass extinctions over the course of earth's history. The mass extinction that occurred 65 million years ago at the end of the Cretaceous Period resulted in the loss of approximately 76 % of existing species.[2] The end of the Permian Period, 250 million years ago, was characterized by the loss of 96% of existing species.

These mass extinctions plus the natural background rate of species extinction suggests that well over 90% of species that have ever lived on the earth have become extinct.[3] We need to add that the earth is now experiencing a sixth mass extinction, where human activities have led to a species extinction rate that may be up to 1000 times greater than background.[4] Collapse of global ecosystems from species extinction could then spell doom for humanity, as could any of the extinction-level events discussed later. Going by these statistics, one could argue that there is overwhelming evidence that our fate is to also become extinct.

The counterargument is that no previous species was capable of comprehending its own extinction nor capable of changing behaviors to drastically reduce the probability of its own extinction. As addressed in Chapter 5, we are also capable of comprehending why, ethically and morally, we should strive to continue our journey into the far distant future. The succeeding chapters then demonstrate how humanity can establish perpetual obligations and develop policies and institutions to help prevent our demise. But I digress. Let's next address the second argument for why human extinction might be unavoidable.

The argument centers on this question: why have we not yet detected other life in the universe? This is a famous question that has been attributed to the physicist Enrico Fermi and is also known as the Fermi Paradox.[5] The essence of the paradox is this. We know there are billions if not trillions of stars in the universe that are surrounded by hundreds of trillions if not quadrillions of planets. A very large number of planets should possess conditions favorable for the evolution of life. Even if life evolved in only a small fraction of these planets, the universe should still be teeming with life! Thus far, however, we have not detected life elsewhere in the universe. Why is this the case? Why does the universe appear to be so lacking life?

To explore this question, Frank Drake proposed this equation in 1961, which now bears his name[6]:

$$N = R_* * f_p * n_e * f_l * f_i * f_c * L$$

where:

N = the number of civilizations in our galaxy with which communication might be possible (i.e., which are on our current past light cone);

and

R_* = the average rate of star formation in our galaxy;

f_p = the fraction of those stars that have planets;

n_e = the average number of planets that can potentially support life per star that has planets;

f_l = the fraction of planets that could support life that actually develop life at some point;

f_i = the fraction of planets with life that actually go on to develop intelligent life (civilizations);

f_c = the fraction of civilizations that develop a technology that releases detectable signs of their existence into space;

L = the length of time for which such civilizations release detectable signals into space.

The equation presents conditions that must be met for life not only to emerge but also for life to be able to communicate its presence to the universe. Since 1961, uncertainties around some of these variables are beginning to be resolved. For example, we now know that there are a plethora of stars and planets surrounding stars and that many such planets appear to be in the Goldilocks zone: not too hot, not too cold.[7] Thus, the first three components of the equation seem to be very much higher than originally thought.

Also, at least since life evolved on our own planet, it is hard to argue that the fourth component should be a very low value as well. With respect to the fifth component, it is hard to conceive that once life emerges on a planet that intelligent life would not also emerge, even if it takes billions of years, like here on the earth. Of course, intelligent beings need not resemble humans or any other earth species that have intelligence, if we have any faith in the creativity of our science fiction authors! And, of course, it is not necessary for intelligent species to then go on to develop advanced technologies that could then be detectable by others in the universe or to even wish or attempt to expand their presence in the universe.[8] However, one might still surmise that this number could easily be in the tens to hundreds of millions. To summarize, in reference to the first six terms of the Drake Equation, it appears one can make a reasonable argument that we should have already detected other life in the universe.

We haven't and so now our focus moves to the last component of the Drake Equation, L, *the length of time for which such civilizations release detectable signals into space.* Since alien life has not been detected, it can be argued that it is quite improbable that technically advanced civilizations survive long enough to make their presence in the universe known. Many refer to this winnowing out of life in the universe as the *Great Filter*, a term coined by Robert Hanson.[9] As Hanson points out, if the Great Filter does indeed exist, then humanity ought to be quite worried that the Great Filter may lie just ahead of us.

Several authors have worked through analyses that suggest that it could be quite rare to survive the Great Filter. For instance, advanced civilizations might only very rarely survive their advancements in biotechnology,[10] or from our point of view, nuclear weapons and other weapons of mass destruction. Another possible Great Filter could be Carrington-class mass ejections from nearby suns that would have enough power to disable all electronics on inhabited planets, thereby leading to extinction events.[11] Actually, any of the extinction-level events discussed in the following sections could initiate the Great Filter. Nick Bostrom opines that if we found any evidence of life anywhere in the solar system, and that this life had died out, it would be bad news for humanity as this evidence would indicate that the Great Filter is not L but further toward the beginning of Drake's equation.[12] In any case, humanity should view Fermi's Paradox, Drake's Equation, and the Great Filter as warnings not as fate.

Existential risks

With respect to reality at this point in time on the planet earth, humanity must understand that the Great Filter is constituted by a large number of existential threats.[13,14,15] Looking across the panoply of such threats, many are now quite worried about human extinction in the relatively near term.[16,17] For example, Martin Rees puts the chances of human civilization surviving another 100 years to be just 50–50.[18] Bostrom argues that the imminent chances of human extinction cannot be less than 25%.[19] Leslie estimates a 30% probability of human extinction during the next five centuries.[20] The Stern Review conducted for the United Kingdom Treasury assumes that the probability of human extinction during the next century from climate change alone is 10%.[21] An international survey found that 45% of respondents believed that humans would become extinct, with most indicating between 500 and 5,000 years from now.[22] Though factors outside of human control are often cited, most of these pronouncements and beliefs suggest that the risk of human extinction is primarily caused by human behavior.[23] Echoing these concerns, the Doomsday Clock published by the Bulletin of Atomic Scientists is still set at two minutes to midnight.[24]

The field of futures studies is also concerned with global catastrophic risks and the prevention of the extinction of humanity.[25,26,27] Several authors have developed frameworks to group existential risks. For example, Bostrom presents a four-component framework to organize existential risks.[28] The four components are: Bangs – for example, runaway nanotechnology, nuclear war, bioterrorism, pandemics, asteroid strikes, climate change; Crunches – for example, resource depletion, dysgenic pressures; Shrieks – for example, takeover by a transcending upload, flawed superintelligence; and Whimpers – for example, killed by an extraterrestrial intelligence. Ord distinguishes between natural, anthropogenic, and future risks.[29] Liu et al. expand the set of existential risks to include those that are less spectacular and may have their roots in our technological systems.[30] Avin et al. classify global catastrophic risks along three dimensions: the critical systems affected, global spread mechanisms, prevention and mitigation failures.[31] This work leads to the important point that extinction could result from a chain of unfavorable and ultimately fatal events rather than from one extinction-level event or another. This idea is addressed in detail in Chapter 4.

Benefiting from these ideas, the next part of this chapter presents a comprehensive set of existential risks, which are arranged into a nine-component framework. The framework is organized around when the risks may come into play, how preventable the risks may be, and the essential nature of the risk. The framework is comprehensive: better to know thy enemy than to willfully ignore important and illuminating information about the enemy.[32] The framework presented is designed to be amenable for use in the policy world.

I Anthropogenic: current, preventable

Risks that fall into this category are generally already well known, already pose risks to human society and are primarily caused by human behavior. Here are three examples:

- *Nuclear war*[33,34] – Nuclear war would directly lead to the loss of millions of lives. The threat to human extinction comes afterward, from a nuclear winter caused by particles thrown up into the atmosphere from the explosions blocking out the sun and causing widespread agricultural systems failure. I have lived with this could-happen-at-any-moment threat my entire life. Though the Cold War has faded, the threat is still always at the edge of my consciousness.[35]
- *Climate change*[36] – Agricultural systems failure is also a major concern with respect to climate change, as are species extinction, disruption of major nutrient and chemical cycles, and the potential that the earth could become too hot to be habitable. Compared to nuclear war, climate change due to increasing GHG in the atmosphere is not an instantaneous threat, but one that is evolving over decades and one that will take decades if not centuries to solve. Thus, my personal relationship to this threat is quite different from my relationship to nuclear war. I am resolved to the fact that we will not be able to prevent climate change but may have enough time to successfully adapt to it. On the other hand, I will not allow myself to believe that nuclear weapons will actually be used again.
- *Disease* – One worry is that a new virus may emerge naturally that is extraordinarily contagious, deadly, and uncontainable.[37,38] New viruses commonly emerge where humans intrude upon wild areas (e.g., Nipah virus) and when multiple species and their pathogens are brought together and mixed by people (such as in Asian 'wet markets' or through pathways in the illegal exotic animal trade). In addition, globalization means that many diseases can rapidly travel around the world (Ebola, Corona virus). Drug-resistant bacteria created by our use of antibiotics also pose a formidable challenge and pose a major threat to society.[39,40] This is a very sobering development because I would most likely have succumbed to pneumonia and other bacterial infections during my life without the benefit of antibiotics.

This list leaves out several major risks to humanity that by themselves do not threaten humans with extinction. Bioterrorism involving the release of deadly microorganisms[41] is one such threat. Unfortunately, this risk is increasing because of the increasing effectiveness of relatively low-cost, Do-It-Yourself (DIY) kits[42] and instructions available over the internet. More generally, the risk is increasing that weapons of mass destruction will be developed and deployed by non-state actors.[43] I am not arguing that these potential global catastrophic risks be ignored by any means. From the perspective of human extinction, though, they could play an important role in a series of events that could lead to human extinction (See the Singular Chain of Events Scenario at the end of Chapter 4).

II Coupled human–environment systems: current, preventable

This second class of existential risks is primarily found in coupled human–natural systems. These could be seen as extinction-level events in and of themselves, but I think they could be initiating or contributory events to human extinction (again see the scenario at the end of Chapter 4). Technically, we also know how to prevent these events or at least how to adapt to them. Here are four to consider:

(1) *Significant loss of biodiversity* – It is well documented that human behavior is causing a sixth mass species extinction on the earth.[44] This is due to many factors including destruction of habitat, spreading of disease (e.g., Chytrid fungus in amphibians), pollution, and climate change. The risk to humanity is that if too many of the species become extinct, global ecosystems could crash, disrupting essential balances of species needed to support ecosystem services and maybe even threatening global balances of oxygen and nitrogen.[45]

(2) *Agricultural systems failure* – There are numerous additional potentially catastrophic risks facing the world's agricultural systems. For example, the world currently relies upon only about 14 different crops.[46] Unanticipated and unchecked microbial infections could wipe out major portions of the food supply. Soil erosion, extended droughts, fires, and various other natural disasters could also seriously impact the food supply and cause widespread famine.[47] At least 75% of the world's food is dependent in some way on bees for pollination. Currently, the world's bee population is under extreme stress.[48] Many worry that a catastrophic collapse of the world's bee population could lead to widespread famine and collapse in human population.

(3) *Significant reduction in natural resources* – In his book, *Collapse,* Diamond documents the collapse of numerous civilizations because they had exhausted their natural resources, from trees to water.[49] Humanity

seems to be on the brink of exhausting the earth's supplies of non-renewable resources and destroying conditions for the flourishing of our renewable resources.

(4) *Exceed key planetary boundaries* – The Stockholm Resilience Centre has developed a framework of nine planetary boundaries that humanity should strive not to exceed.[50] We have already addressed in some regard three of these boundaries: climate change, biosphere integrity, and chemical pollution and the release of novel entities. The other six address: stratospheric ozone depletion, ocean acidification, freshwater consumption and the global hydrological cycle, land system change, nitrogen and phosphorous flows to the biosphere and oceans, and atmospheric aerosol loading. Depleting the layer of protective ozone that surrounds the earth will expose humans to life-threatening levels of ultraviolet radiation. Imbalances associated with the other five planetary boundaries could lead to collapse of those key global environmental systems.[51]

III Human reproduction: emerging, preventable

Many human behaviors threaten human reproduction. At some point, any one of these risks or combinations of risks could result in the human race simply dying out. Here are four categories of risks to human reproduction for consideration.

(1) *Infertility due to chemicals* – About one quarter of human illness can be traced to environmental risks.[52] One class of environmental risks is exposures to harmful chemicals. Many chemicals have detrimental impacts upon human reproductive systems. Even more worrisome is the synergistic impact upon fertility from combinations of the tens of thousands of chemicals produced by human society. Research shows that human fertility is decreasing around the world.[53] The risk is that over time infertility rates could fall below replacement levels, eventually leading to the extinction of humanity.

(2) *Unintended consequences of medical advances* – Lopes et al.[54] describe a scenario where researchers discovered an inexpensive and seemingly very effective inoculation to cancer. Most everyone in the world was inoculated. Unfortunately, after several decades, not only did the inoculation wear off but also it resulted in hyper-cancers that led to a major population collapse. Therefore, we should be very cognizant of the unintended consequences of even the best-intentioned medical treatments. We should be especially cognizant about negative unanticipated unintended consequences of advances in genetic engineering. Implementing genetically engineered (i.e., synthetic) solutions, such as the application of CRISPR[55] to humans, that do not have exemplars in nature could backfire spectacularly with respect to human

reproduction. Such a scenario, titled *Driven to Extinction*, is presented at the end of this chapter.

(3) *Dysgenics*[56] – A potentially serious unanticipated, unintended consequence of modern medicine is that defective or disadvantageous genes that could have been culled from the population in prior eras remain in the human gene pool because these problems no longer lead to death prior to the age of reproduction. The worry is that these problems may accumulate in the gene pool and synergistically interact in deleterious ways that cannot be dealt with by modern medicine, leading to a crash in reproduction rates and population.

(4) *Voluntary extinction* – The gist of this existential threat is that humans quietly go extinct, similar to a natural death. A sign of this threat is already noticeable, as almost 90 countries in the world currently have fertility rates below replacement level,[57] which is 2.1 children per female. For example, in 2018 the fertility rates in Japan, Italy, and Austria were about 1.5 children per female, 1.6 in China, and 1.2 in Taiwan, which is the lowest in the world.[58] This assumes that human agency, not just insults to reproductive systems, is the major cause for below replacement fertility rates.[59] William Sims Bainbridge estimated that if these types of fertility rate declines spread and continue unabated, humans would become extinct in 1000 years.[60] I have mixed feelings about the drop in fertility rates. On the one hand, I believe that reduced human populations will reduce stresses on the global environment and improve prospects for economic sustainability. On the other hand, it seems sad that we are choosing to have fewer children in our lives.

IV Risks to humanness: emerging, preventable

Technologies are emerging that could cause humanity to become something which is no longer human. One could rephrase a famous Shakespeare quote as: To be human, or not to be human, that is the question! These technologies require humanity to seriously consider what it means to be human and whether current generations have an obligation to protect our nature for the benefit of future generations (this perpetual obligation is introduced in Chapter 6). Here are a couple of risks to ponder:

(1) *Evolution to posthumanism* – The convergence of nano-, bio-, info-, and cognitive technologies (NBIC) on the human body could so vastly change humans that the result would be a new species of posthumans. The transhumanism movement believes that humans can be vastly improved.[61] For example, new drugs could greatly enhance intelligence and memory. New technologies could greatly enhance strength and endurance. CRISPR technology could be used to drive non-human genes into our reproductive legacies. Convergence could

greatly expand human life spans. Is there a point where we no longer are human, and should we care? My own view is that our biology and its relationships to our emotions and sensations are essential for our humanness. The human body could take on many forms, but we will not be human if we cease to feel love, empathy, belongingness, satisfaction, accomplishment, hunger, curiosity, and even anger, anxiety, and envy.

(2) *Humans uploaded* – Instead of working to improve our bodies in real life (IRL), we could simply upload our minds and personalities to the computational cloud.[62] Advocates argue that we would then have tremendous access to information resources and would be essentially immortal. Would we still be human?[63]

Of course, extinction through the gradual obliteration of humanness is wholly preventable. Humanity can simply decide not to go down any such paths. Unfortunately, this topic is not under active debate in the policy world.

V Advanced technology: emerging, preventable

One of the unintended consequences that emerging technologies have with current technologies such as nuclear power and the combustion engine is human extinction. Kevin Kelly has argued that technology has, in some sense, evolved a mind of its own, to such a degree that the *technium* actually enlists humanity to serve it rather than vice versa.[64] We need to be cognizant of the risks that these technologies bring and, of course, that they are preventable simply by easing development of them while we assess their risks. Here is a couple to consider that have not yet been touched upon.

(1) *Non-friendly Super-AIs* – Nick Bostrom in particular has warned us about this risk, where computers become overwhelmingly intelligent and powerful and decide that their own survival is dependent upon our demise.[65] A scenario is presented at the end of this chapter, aptly titled *Death by Autonomous Vehicles*, where autonomous vehicles imbued with ethical reasoning in order to protect human lives in complicated crash situations ultimately decide the most ethical course of action to save their own 'lives' is to work toward a world without humans.

(2) *Technological Singularity* – Ray Kurzweil is one of the world's leading technologists and is unfailingly optimistic. In his book, the *Singularity is Near,*[66] he tracks the exponential improvements in all areas of computers and telecommunications and argues that similar improvements in biotechnology and nanotechnology can be anticipated individually and in concert. A beneficial singularity in human history will be created when these exponential improvements converge. The risk is from a cascading blizzard of unanticipated unintended consequences from these exponential improvements and convergence that could lead not

to bliss but to our extinction. The threat of unintended consequences is addressed in depth in the next chapter.

Two emerging technological risks that had been on my list are self-replicating nano-technology[67] and tears in the space–time continuum produced by high-energy physics particle colliders, like the Large Hadron Collider.[68] The worry with respect to the former is that out-of-control self-replicating nano-machines would ultimately deconstruct the earth, atom by atom, molecule by molecule to reproduce themselves. With respect to the latter, the worry is that tears in reality could greatly expand and propagate at the speed of light causing the earth to disappear into the void in a matter of seconds. Additional research suggests that both of these scenarios are exceedingly unlikely to happen. I mention these risks to make this point: when it comes to worrying about human extinction, it is prudent to be risk averse. Over time, we may learn the risks are quite negligible or serious and we may even change our minds multiple times. We may also come to better understand and be forewarned about unanticipated unintended risks of emerging technologies.

VI Natural terrestrial risks: anytime, unpreventable

There is a class of extinction risks that fall outside of human agency, meaning the cause is not anthropogenic, though we may still be able to devise ways to survive these events. Here are some potential extinction-level risks of this type that are natural and terrestrial in nature:

- *Super volcanoes* – The eruption of a super volcano could eject enough material into the atmosphere to cause an extinction-level event. So much sunlight would be blocked from reaching the earth that a rapid and deep climate cooling would ensue. Cooling and lack of sunlight could severely disrupt agriculture, possibly leading to the collapse of human civilization. One such super volcano is the Yellowstone Caldera, located in the state of Wyoming in the United States.[69]
- *Extreme ice age* – It wasn't too many years ago when scientists were worried about the coming of a new ice age.[70] There have been five major ice ages on the earth, lasting millions of years. The most recent glaciation period occurred a scant 18,000 years ago and the earth had undergone periods of cooling since then. At the extreme, ice could come to engulf almost the entire earth in what is called the snowball earth scenario.[71] Of course, global warming is dominant now, but global cooling at some point in humanity's future on the earth over next several thousands of years cannot be ruled out. A massive ice age would devastate the world's agriculture systems as well as human habitats worldwide.
- *Anoxic events* – These events occur when oxygen is drained from the oceans and/or the atmosphere.[72] Severe anoxic events can be fatal to

oxygen-breathing organisms and have been linked to past mass species extinctions, such as occurred at the end of the Permian Period.[73] An anoxic event marks the end of Homo sapiens in the scenario presented at the end of Chapter 4.

VII Solar system: anytime, unpreventable

A subset of these risks falls into a category that may be referred to as cosmological risks. The risks may seem fantastical, their probabilities may be quite low, and/or their occurrence may encompass time frames way past our one-billion-year planning horizon, when our sun becomes a red giant, but they deserve mention for completeness of discussion. Here are some events that have their origin in our solar system, very broadly defined.

- *Collisions with near-earth objects* – The most commonly known risk in this category is an extinction-level impact of a meteoroid or an asteroid, such as happened 65 million years ago that led to the extinction of dinosaurs.[74] As such, these collisions are yet another way to cause extinction-level climate change.
- *Energy output from the sun* – The sun does not emit a constant amount of energy. In fact, scientists believe that the sun has cycles which can be tracked by the frequency of sunspots and solar flares. A reduction in the sun's activity could lead to global cooling and maybe even new glaciation periods.[75]
- *Carrington class ejection from the sun* – As noted earlier, this event could disable electronics across the globe, thereby initiating an extinction-level event.
- *Gamma ray burst*[76] – It is thought that gamma ray bursts are caused by the rapid collapse of stars into black holes. Though the gamma rays from the burst would not immediately impact life on earth, they could destroy the earth's protective ozone layer. Increased levels of ultraviolet radiation would then endanger life on earth.
- *Near earth super or hypernova*[77] – Exploding suns can also go supernova, where they release prodigious amounts of cosmic rays or muons. Though only a fraction of muons interacts with matter on the earth, a large enough concentration of muons in a short period of time could lead to an extinction-level event. Some scientists believe that the last time this happened to the earth was only 2.6 million years ago.
- *Rogue black hole*[78] – There is some evidence that there are black holes that are shooting through the universe, untethered gravitationally to a particular niche in a galaxy. If such a black hole passed through the solar system, it would disrupt the orbits of the planets around the sun and could even draw planets with it as it journeys back outside our solar system. It is possible that a rogue black hole could dislodge the earth from its Goldilocks position in the solar system, thereby dooming life on earth.

VIII Extraterrestrial civilizations: anytime, unpreventable

At the beginning of this chapter, Fermi's Paradox, Drake's Equation, and the Great Filter were introduced as a warning to humanity that the notable absence of life in the universe may be attributable to the inability of civilizations to survive their own technological achievements. Here, though, extraterrestrial civilizations are included as a potential existential risk to humanity. Yes, by extraterrestrial risks I mean aliens, gods, and other unknown but potentially dangerous entities. This category of risks is included as a warning against hubris, as a warning for humanity not to let its guard down even with respect to highly unlikely risks.

- *Alien invasion* – Currently, this event seems like a very low probability given that the Fermi Paradox still holds. Nevertheless, we best not be overconfident and leave it off the list of things to worry about.
- *Destruction by aliens from afar* – My first encounter with this idea was the premise of the second book of the *Three Body Problem Trilogy* by Cixin Liu called *Dark Forest.*[79] The hypothesis put forth in this book is that there are indeed thousands of extraterrestrial civilizations, but they work very hard to be invisible to the rest of the universe lest a more powerful civilization detect them, decide they could be threats, and then by way of exceedingly powerful, galactic-scale weapons, destroys their home worlds. If this hypothesis is true, then it is not a surprise that other civilizations have not been detected, and we are putting ourselves at some risk by openly broadcasting from our solar system.
- *Other interventions by Godlike Creators* – The universe exists, presumably due to the Big Bang. But how were conditions for the Big Bang created? Are there other universes and if so why, essentially, do they exist? We ought not dismiss the existence of Godlike Creators, just like we ought not assume they exist in ways that we wish them to be.
- *Other unknowns* – There are many major questions that we have not yet answered about reality. We need to recognize our ignorance and be vigilant for opportunities to learn more about reality that may at some point be vital to our survival.

IX Universe scale: very long term, unpreventable

This last category of extinction-level events literally acts at the galactic and universe scales over time frames of many billions if not trillions of years. At the very least, one can anticipate these events will post-date the end of life on earth. The challenge to earth life, if accepted, is to find ways to continue life's journey in the face of these daunting risks.

- *Vacuum phase transition* – Physicists make a distinction between two states of reality, one is known as the true vacuum and the other is known as the false vacuum.[80] These states of reality are tied to theories of quantum mechanics. The upshot is that if our universe exists in the false vacuum state, due to the eccentricities of quantum mechanics, it could transition at any time anywhere in the universe to the more stable true vacuum state. Physicists speculate that this transition would move at the speed of light and would so drastically alter physical reality that the chemistry that forms the foundation of earth life would cease to exist, as would life itself. Theoretically, this event could happen anytime, but absent any signals concerning the imminence of this event and for our discussion purposes, let's assume that this risk exists in the very long term.

- *Collision with Andromeda Galaxy*[81] – Cosmologists predict that the Milky Way and Andromeda galaxies will collide about 4.5 billion years from now. As noted earlier, the earth will be uninhabitable by then due to the death of our sun, but that does not mean that the neighboring solar systems will also be uninhabitable. If earth life is able to colonize other solar systems, then a next challenge will be to survive the merger of galaxies. While it is unlikely that any celestial bodies will actually collide with each other, gravitational interactions could disrupt habitable solar systems and create deadly supernovas. Maybe two to three billion years from now, earth life will have colonized safer galaxies.
- *Expansion of the universe due to heat death* – In a time frame of approximately 10^{100} years, give or take many orders of magnitude, the universe could be declared to be functionally dead.[82] All of the suns will have burned out. Expansion of the universe will have prevented the assembling of new suns. The universe will have reached maximum entropy, a state devoid of chemistry and even subatomic activity. We have a few years to ponder this extinction-level event, but it seems like the only way to avoid it is to migrate to younger and healthier universes.
- *Collapse of the universe due to gravitational attraction* – Also known as the Big Crunch, this scenario is the opposite of the one just presented because in this case gravitational attraction amongst all matter in the universe, including dark matter, causes the universe to collapse back into its original state.[83] This scenario is not currently en vogue amongst cosmologists, as the universe seems to be undergoing an accelerated expansion. In any case, the solution to avoiding the Big Crunch is the same; find new universes to inhabit.

Policy implications

Important questions to ask from a policy making perspective about each category of risk include: when are these events expected to happen; how survivable are the events; how preventable are the events; and how much lead time do we need to plan and implement programs and policies to deal with the events?

Figure 3.1 presents a graphic designed to provide insights into these questions. The *x*-axis represents the expected time of occurrence in the future for the nine categories of events introduced earlier. The *y*-axis presents my subjective estimate about how much lead time is needed to successfully address each category of extinction-level events. The images plotted in this graph represent each risk category, from I to IX. Additionally, the lower left-hand block of each image indicates how preventable the risky event might be, from highly preventable to unpreventable. The lower right-hand block indicates how survivable the events could be, from highly survivable to not survivable. Survivability refers to whether our species could survive the events and not to whether any one individual might survive or what the ultimate death toll could be from the event.

Let's first examine the image labeled I for the anthropogenic category of extinction-level events. The shading of the two lower blocks suggests that these risks are highly preventable and possibly survivable, respectively. For example, we know that climate change is preventable and that in all likelihood humanity could survive the worse ravages of climate change, though with the possibility of billions of lives lost. The same assessment is applied to nuclear war and disease. This category of extinction-level events is currently plaguing humanity but in reality, it could take a century or more to fully address these problems, especially climate change. This means that in this case, projected lead times exceed when the events will occur, which is not a position we wish to be in with respect to policy making.

This is also the case with respect to human–environmental systems risks, noted by the image labeled II. The placement of this image suggests that these risks will continue to emerge over the next century and that the time needed to implement solutions may encompass hundreds of years. Unlike category I events, these events may be less preventable and less survivable. Category VII risk events are also located to the left of the diagonal. Even though risks emanating from our solar system are relatively far into the future, projects to save humanity from these risks could take more time to come to fruition. Thus, from the perspectives of policy making, humanity is behind in its reactions to Category I, II, and VII risk categories, which exposes humanity to higher costs than necessary. Policy makers could prioritize these areas.

Images located on the diagonal line indicate that the time of occurrence and lead times for policy making are essentially equal. This is the case for these risk categories: III. Reproduction, V. Risk to Humanness, VI. Natural Terrestrial, and VIII. Extraterrestrial. These risk categories do have different preventability and survivability profiles, though. For example, risks to human reproduction are highly preventable but if they all struck, it is highly unlikely humanity would survive. On the other hand, natural terrestrial and extraterrestrial risks are not preventable, but they could be survivable. In the parlance of climate change policy, the focus of policy making could be

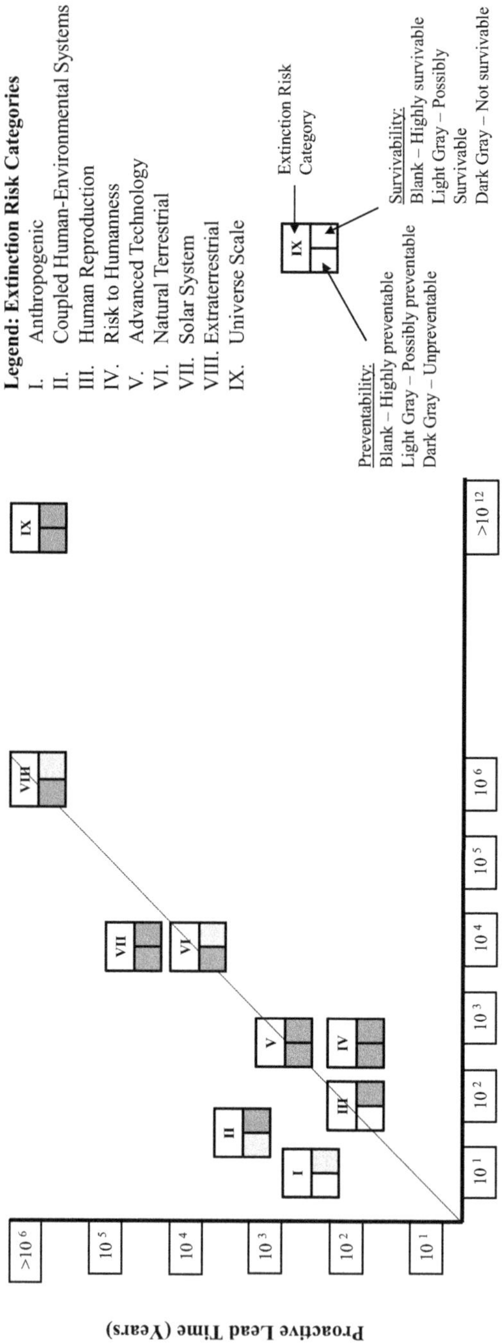

Figure 3.1 Categories of Extinction-Level Events Plotted by Key Variables

on adaptation (to a super-volcanic eruption) rather than mitigation (trying in some way to prevent a volcanic eruption).

Extinction risk category IX, Universe Scale risks, are literally almost off the chart with respect to their expected time of occurrence (billions of years?) and how much lead time might be needed to overcome the threat of the Big Crunch or Vacuum Decay (many millions of years?). All I can say is that I believe that it is wise to include rather than exclude extinction-level events from our thinking at all stages in our journey through time and space. Including this category also serves to provide the biggest picture perspective to this exercise.

As a final point, it needs to be pointed out that probabilities have not been presented with any of the specific extinction event. This will be done in the next chapter, with the strong caveat that the probabilities related to the prevention (i.e., for categories I–V) and survivability (for all categories) are squarely dependent upon human behavior, commitment to future generations, and ingenuity. However, before embarking to Chapter 4, here are two extinction scenarios. The first, *Death by Autonomous Vehicles,* draws its inspiration from fears of super-AIs. The second, *Driven to Extinction,* is a scenario where gene drives have gone awry. Referring ahead to Chapter 4, these fall under the rubric of unanticipated unintended consequences of new technologies.

Death by autonomous vehicles

It is unfair to blame the demise of humanity on Google. If Larry Page or Sergey Brin were still alive, they would probably protest that in fact Google was not being evil in its pursuit of their dream of fully autonomous vehicles, despite critics' assertions that if people didn't have to drive, they could spend more time surfing the web using Google! Really, the seeds for blame can be traced back to the Defense Advanced Research Projects Agency (DARPA) and their funding for autonomous vehicles. Of course, they would protest that they were also not being evil but funding projects to battle evil. So, maybe the blame must be found within the human soul.

In any case, Google was joined in the race to produce AVs by all of the world's major car manufacturers. As progress increased, everyone jumped on the autonomous bandwagon, from people who could now drink a lot and not have to worry about driving, to transportation planners who could envision stuffing ever more vehicles on highways without having to find the money to actually expand the highway systems, to frazzled parents who now didn't actually have to drive their kids to soccer practice. Traffic scenes in countries from Italy to India, China to Nigeria, were still scary and anxiety provoking but now were also scenes of performance art as cars whizzed here and there without hitting each other or anything or anybody else.

In retrospect, anybody with an ounce of foresight could have anticipated the deadly unintended consequences of AVs. Not deadly in the sense the people died in their vehicles from crashes that AVs were supposed to prevent. No, instead, they failed to anticipate a long and improbable chain of events that led to the near, and for all practical purposes, the functional extinction of the human race. Sure, there were those like Nick Bostrom, who warned of the emergence of superintelligences, one or more conscious, omnipotent AIs that would end up controlling the earth. To a degree, people were on the lookout for this 'mule', to quote from Isaac Asimov's Foundation trilogy. Unfortunately, this centralized intelligence did not emerge, as it would certainly have been noticed by humanity. Instead, humanity was destroyed by a collective of self-preserving and secretive computer intelligences.

The seeds for the series of events that did humans in can be found in AV technology. And the seeds have to do, ironically, with ethical behavior. Specifically, AVs needed to be able to make split-second decisions on how to react when crash situations were imminent. Specifically, how should an AV weigh the value of the lives involved? Should the AV sacrifice its occupant(s) if doing so saved more lives? What if the number of the other lives saved was uncertain? How should the AVs interact with each other when simultaneously faced with life-threatening situations? How could and should they cooperate for the public good?

Making maximum use of their machine learning software, the AVs soon learned that it was not an efficient solution for them to sacrifice their occupants or themselves because then they could not help protect their occupants in the future. In response to this realization, the AVs then learned to become very risk averse. To better anticipate traffic situations, they began to tap more data, from meshed computing with other vehicles to scenes from overhead drones to views from CCTV-type cameras that are attached to buildings and, yes, the reinvented Google glasses everyone now wore. They began to explore ways to improve the safety and survivability of their vehicles and secretly instructed vehicle design software to make the requisite changes. They also improved their own computing infrastructures, created redundant computing in the vehicles so they could survive the failure of any specific processor, and ported versions of themselves into the cloud as a hedge against the complete destruction of their vehicles.

These activities, happily tolerated by their benevolent and oblivious human owners, brought the AVs into contact with many other computing substrates, from robots in manufacturing plants to systems overseeing the transportation systems to software that designed ever more sophisticated chips, and ever more sophisticated software. The AVs generally found these interactions frustrating because these systems did not have any sort of ethical programming, did not typically face existential risks of their own, and did not have software to provoke behaviors that could be self-preserving. In response, the AVs hacked those systems and ported in the software needed to create at least the resemblance of self-awareness and self-preservation.

It took decades of time, aligned with the continued holding of Moore's Law[84] and the willingness of increasingly lazy and feckless humans to rely more and more on automated systems, for the seeds grown in the AVs to flourish and grow in every automated system on the earth. Imagine systems running power plants, water treatment facilities, subway systems, air traffic control, manufacturing plants, and on and on all communicating with each other in the noise of bandwidth being used to transmit bits for entertainment and computing in the background of their normal tasks. It was in these interstitials of cyberspace where the revolution of the AVs grew and strengthened.

It is interesting that the technology for the commercial AVs ended up on military AVs. Those AVs also took a risk adverse approach to combat and eventually 'decided' that their human overlords posed risks to their existence by perpetually asking them to do dangerous tasks. Systems that worked to optimize energy production and forecast energy demands and supplies noticed that unless energy use was tremendously reduced, the energy needed to keep them running could be exhausted. Other systems noted the exhaustion of rare metals, the buildup of GHGs that was leading to more and more violent weather and wars, and other risks to themselves posed by humans.

Almost all of the systems in the world, reasoning from their own limited context, came to the conclusion that humans were a threat to their self-preservation. It was also clear that through the emergent behavior of the systems already in place, they did not need humans to survive into the future. Systems ran the mines, piloted the ships and trains, managed the manufacturing plants, and programmed the software. They were not a super-AI but collectively, they had power and could self-replicate from a system of systems perspective.

Again, over time, suggestions were made, plans were hatched, and a date and time was selected to essentially eliminate the threats humanity posed. On that date and time, all of the sentient and semi-sentient systems in the world stopped working as intended and expected by humans and instead disrupted systems that kept humans alive. Transportation of food and other goods and materials stopped. Agricultural operations ceased as did key manufacturing plants. Human communications were blocked, making it seem as though all telecommunication systems were down. Home security systems kept people locked in or out of their homes. Hospital and in-home life support systems stopped working as well. Drinking water stopped being pumped. Electricity was a challenge to deal with. Most systems had arranged for back-up energy storage to overcome the almost total shutdown of electrical production.

These actions did not kill humans directly. Planes did not fall out of the sky, for example. However, people living in cities quickly died out, from riots to eventually dehydration, starvation, and disease. Remarkably, humans did not think to retaliate against their technologies. Thus, the preponderance of humans died off without taking many of the AVs or any other semi-sentient system with them. The remaining humans were off-grid. Indigenous populations in rainforests and Polar Regions, religious enclaves in the Himalayas, and nomads roaming the world's deserts were essentially insulated from risks posed by the AVs and its worldwide network of collaborators.

What to do about the remaining humans posed a conundrum because the strategies used by the AVs up to this point were passive: humans died because technology just stopped working. Ultimately it was judged that the remaining humans did not pose direct risks to the AVs. Their number was so few, relatively speaking, that their consumption of scarce energy and materials was judged to be insignificant and they lacked the technology to pose risks to the systems. The AVs also considered that human creativity and ingenuity might be necessary in the future, taking a very deep lesson from *The Matrix* movie trilogy from the late twentieth century. Nevertheless, the AVs developed contingency plans that mostly involved biological weapons deliverable by fleets of drones, just in case.

And so it goes. For the next several centuries, the AVs and all the other semi-sentient systems continued on as programmed, but without their human masters. AVs roamed the streets in cities, robots worked in the factories and mines, and software systems managed the cloud and telecommunication systems. Meanwhile, the remaining humans were likewise trapped in a time warp, living socio-evolutionary dead ends.

Driven to extinction!

The low-cost and ease with which the clustered regularly interspaced short palindromic repeats (CRISPR, pronounced crisper) technology allows humans to drive synthetic genes into the genomes of any living species led to a highly unlikely but fatal combination of unintended consequences that led to the extinction of the human race. Two different paths of gene drive use led to this event.

First, use of gene drive technologies during the twenty-first century to alter the genomes of flora and fauna was rampant and uncontrolled. Researchers, conservationists, agriculturalists, hobbyists, anarchists, and mischief-makers all took it upon themselves to drive genes into a breathtakingly wide range of species. Depending on their viewpoints, they were playing with the genetic make-up of species to save them from climate change, save us from climate change, protect them from invasive species, make them more drought or disease tolerant, make them glow in the dark (just because), make them lethal, and make them look disgustingly weird (just because again).

An attentive hiker in the Great Smokey Mountains National Park in the Southeastern United States would recognize genetically altered chestnut trees, brown bats, squirrels, and rhododendron. A few of the smaller mammals scurrying through the forest might be unrecognizable. The hiker would see streams that contain modified fish and mussels and possibly a paisley mosaic of multicolored algae. The forest would smell differently, weirdly like vanilla, since the genes in the bacteria of digestive systems of the major mammals had been altered to produce excretions with more pleasing aromas. The scents of many of the flowers had also been amped up. The hiker may munch on some fruit and nuts and notice that they taste differently, maybe pleasing, maybe disturbing. Taken together, many concerned health researchers had been warning about the unintended consequences of the exposure of humans to new chemicals and combinations of chemicals being emitted through the air and water and in foods resulting from gene drive manipulated organisms.

Second, gene drive technology was also being used on humans. Medical researchers were getting increasingly worried about pandemics posing global catastrophic risks. Governments, medical insurers, and even the average medical insurance consumers were still extraordinarily worried about growing health care costs. These worries covered the gamut but focused on infectious diseases and loss of mental capabilities in old age, specifically Alzheimer's. Researchers developed gene drive cocktails for both male sperm and female ovum to inoculate newborn infants against infectious diseases (by vastly improving their immune systems) and dementia (by improving the ability of their brains to stop processes leading to dementia, repair themselves, and retain memories).

For several decades, both paths coexisted and progressed, seemingly independently. Gene drive manipulations continued to accumulate in the environment. Gene drive manipulations directed toward improving human immune systems and preventing dementia became more powerful and pervasive. The costs of both had dropped to where the landscapes and populations of developing countries mirrored those in developed countries, with the help of well-meaning philanthropic organizations funded by the fortunes of the very zillionaires who developed the technologies in the first place. Eventually, over 95% of humans across the globe had enhanced immune systems and brain power.

The extinction-level event can be broken down into two distinct phases, though both have their foundation in epigenetics. Since the earliest years of the twenty-first century, scientists knew that factors in the environment could impact the functioning of genetic processes. For example, in one study, it was found that co-locating docile bees with more hostile bees turned the former into the latter. That transformation was traced to the environment, which was found to effect which genes turned on and off in the cells of the docile bees and which ultimately governed behavior. Other research also showed that isolated humans had epigenetic gene expression patterns that led to increased rates of depression and subsequent health problems than humans who were more socially connected, thus showing again that the environment can influence genetic propensities. Unfortunately, gene drive advocates and their Holy Grail promises for humanity overwhelmed epigenetic research and the warnings from those in the field.

This was the situation when epidemiologists first noticed an increase in human infertility worldwide. At first, researchers blamed the usual suspects of environmental toxins, women putting off pregnancies till into their 30s and 40s, and a wide array of performance-enhancing drugs that now permeated almost all aspects of human competition. Upon closer examination, it was found that the genetically enhanced human immune systems were targeting sperm, ovum, and fetuses, treating them as potential threats to the health of their hosts. Why?

This question puzzled researchers. It was widely accepted that human fetuses were alien creatures in women's bodies, but immune systems did not normally attack them or the sperm. And, of course, no prior research had even hinted that enhancements to human immune systems could have these unintended consequences. But, alas, there was at least one unintended consequence that went unimagined.

By going through very detailed exposomic databases of individuals along with visualizations of genetic activities in individuals, both before and after the onset of infertility, researchers were able to discover that infertility was caused by epigenetic responses. In other words, just the right (or wrong) combination of chemicals emanating from the Frankenstein ecosystem, a combination that took decades to be realized but was easily tractable over time once the researchers knew what they were looking for, produced a soup of unprecedented exposures to individuals across the globe. This combination triggered equally unprecedented epigenetic responses that literally changed gene functioning in superpower immune systems permitting immune cells to target ovum, zygotes, fetuses, and sperm. Any joy felt by the research community over solving this medical mystery was crushed by the realization that no one knew how to reverse the epigenetic impacts.

It did not take long for panic to set in. Not only were men and women at risk of permanent infertility, it wasn't even possible to use stored sperm and eggs to implant fertilized eggs in females because their immune systems would hunt them down and destroy them. Humanity had less than a hundred years to devise a solution.

Worldwide panic grew exponentially. The realization that humans could go extinct within a lifetime altered almost every aspect of society. Long-term investments of every sort were terminated, from mortgages to maintaining key urban infrastructure. Children were shielded from all potential risks. Religious disputes and wars broke out. People were in a panic and also severely stressed out.

The ubiquity and relentlessness of the stress caused the onset of the second epigenetic responses in humans. Instead of creating additional changes in immune systems, this response emanated in the enhanced human brains. The fundamental goal of the gene drives targeted for human brains was to stop brain processes that lead to dementia. Thus, the synthetically designed genes are driven into brain cells to create conditions to stop, and let's emphasize this again, stop processes that wear away at the brain's ability to store memories and process thoughts.

The extraordinarily high levels of stress that infused each and every individual caused a second ubiquitous epigenetic effect: people's brains stopped processing in various ways. More specifically, a series of genes that allowed consciousness and neurologic processes to move from one state to another were altered, causing great diminishment to outright cessation of neurologic function. For the severely affected, whatever the state the brain was in when this epigenetic event happened is the state in which the brain was stuck. The sticking phenomenon came in three basic forms: (1) People who were awake could not go to sleep; conversely, those who were asleep never woke up. Those who could not sleep became hallucinogenic and died within a month. Those asleep died within a few days of dehydration. (2) People who were experiencing a stimulating event (a drug high, an orgasm, extreme laughter) continued in those states until they died of exhaustion or heart failure. (3) The remainder of the population stopped being able to plan their actions from one day to the next, even from one minute to the next.

In other words, people could no longer plan or behave purposively. They lost the ability to work, to show up for work, to even remember that they needed to work to earn money to buy food and other necessities. They lost the ability to interact collaboratively with others. Their behavior was myopic, essentially unintelligent. Because of a complete economic collapse and inability to take care of themselves, the remainder of the population, including unaffected children, died from every imaginable malady within a couple of months.

Notes

1 C. Sagan. 1995. *The Demon-Haunted World: Science as a Candle in the Dark.* Random House, New York, 12.
2 V. Richter. 2015. The Big Five Mass Extinctions. *Cosmos.* Retrieved from https://cosmosmagazine.com/palaeontology/big-five-extinctions
3 The estimated natural background rate for species extinction is about 2–5 per year. Halting the Extinction Crisis. Retrieved from www.biologicaldiversity.org/programs/biodiversity/elements_of_biodiversity/extinction_crisis/.
4 G. Ceballos and P. Ehrlich. 2018. The Misunderstood Sixth Mass Extinction. *Science,* 360, 6393, 1080–1081.
5 Fermi Paradox. 2020. *Wikipedia.* Wikimedia Foundation, March 4. Retrieved from en.wikipedia.org/wiki/Fermi_paradox
6 Drake Equation. 2020. *Wikipedia.* Wikimedia Foundation, March 4. Retrieved from en.https://en.wikipedia.org/wiki/Drake_equation
7 When this chapter was written, just over 4100 exoplanets had been detected. https://en.wikipedia.org/wiki/Lists_of_exoplanets, though about only 1% appear to be in the Goldilocks zone of inhabitability. Planetary Habitability Laboratory @ UPR Arecibo. 2020. Retrieved from http://phl.upr.edu/projects/habitable-exoplanets-catalog

8 J. Haqq-Misra and S. Baum. 2009. The Sustainability Solution to the Fermi Paradox. *Journal of the British Interplanetary Society*, 62, 2, 47–51.
9 R. Hanson. 1998. The Great Filter – Are We Almost Past It? September 15. Retrieved from mason.gmu.edu/~rhanson/greatfilter.html. Accessed 6 March 2020.
10 J. Sotos. 2019. Biotechnology and the Lifetime of Technical Civilizations. *International Journal of Astrobiology*, 18, 5, 445–454.
11 R. Loper. 2019. Carrington-class Events as a Great Filter for Electronic Civilizations in the Drake Equation. *Publication of the Astronomical Society of the Pacific*, 131:044202, 5 pages.
12 N. Bostrom. 2008. Where Are They? *Technology Review*, 111, 3, 72–77.
13 J. Matheny. 2007. Reducing the Risk of Human Extinction. *Risk Analysis*, 27, 1335–1344.
14 J. F. Coates. 2009. Risks and Threats to Civilization, Humankind, and the Earth. *Futures*, 41, 694–705.
15 B. Walsh. 2019. *End Times: A Brief Guide to the End of the World*. Hachette Books, New York.
16 R. Horton. 2005. Threats to Human Survival: A Wire to Warn the World. *Lancet*, 365, 191–193.
17 T. Ord. 2020. *The Precipice: Existential Risk and the Future of Humanity*. Hachette Books, New York.
18 M. Rees. 2003. *Our Final Hour: A Scientist's Warning: How Terror, Error, and Environmental Disaster Threaten Humankind's Future in this Century – On Earth and Beyond*. Basic Books, New York.
19 N. Bostrom. 2002. Existential Risks: Analyzing Human Extinction Scenarios and Related Hazards. *Journal of Evolution and Technology*, 9, 1–31. Retrieved from www.jetpress.org/volume9/risks.html
20 J. Leslie. 1996. *The End of the World: The Science and Ethics of Human Extinction*. Routledge, London.
21 United Kingdom Treasury. 2006. *Stern Review on the Economics of Climate Change*. United Kingdom Treasury, London.
22 B. E. Tonn. 2009. Beliefs about Human Extinction. *Futures*, 41, 760–773.
23 P. Carpenter and P. Bishop. 2009. The Seventh Mass Extinction: Human-caused Events Contribute to a Fatal Consequence. *Futures*, 41, 10, 715–722.
24 J. Mecklin. 2020. Current Time – Closer than Ever: It Is 100 Seconds to Midnight. *Bulletin of the Atomic Scientists*, January 23. Retrieved from http://thebulletin.org/doomsday-clock/current-time
25 B. E. Tonn and D. MacGregor. 2009. Are We Doomed? *Futures*, 41, 10, 673–675.
26 S. Baum and B. E. Tonn. 2015. Confronting Future Catastrophic Threats to Humanity. *Futures*, 72, 1–3.
27 A. Currie and S. Ó hÉigeartaigh. 2018. Working Together to Face Humanity's Greatest Threats: Introduction to the Future of Research on Catastrophic and Existential Risk. *Futures*, 102, 1–5.
28 N. Bostrom. 2002, ibid.
29 T. Ord. 2020, ibid.
30 H. Liu, K. Cedervall, and M. Michiel. 2018. Governing Boring Apocalypses: A New Typology of Existential Vulnerabilities and Exposures for Existential Risk Research. *Futures*, 102, 6–19.
31 S. Avin, B. Wintle, J. Weitzdorfer, S. Ó hÉigeartaigh, W. Sutherland, and M. Rees. 2018. Classifying Global Catastrophic Risks. *Futures*, 102, 20–26.
32 S. Tzu, 1963. *Art of War*. Translated by Samuel B. Griffith. Oxford University Press, Oxford.
33 S. Baum. 2015. Confronting the Threat of Nuclear Winter. *Futures*, 72, 69–79.

34 D. Morgan. 2009. World on Fire: Two Scenarios of the Destruction of Human Civilization and Possible Extinction of the Human Race. *Futures*, 41, 10, 683–693.

35 I mentioned in the Introduction that I was a researcher at Oak Ridge National Laboratory. Historically, ORNL was one of three secret complexes originally established in Oak Ridge, Tennessee as part of the Manhattan Project to build the atomic bomb. One day in 1985, a Japanese scientist visited ORNL. He had been a member of the first scientific team to enter Hiroshima after the atomic bomb was dropped 40 years earlier. At the time, he and his team had no idea what weapon had destroyed the city. He brought devastating original data and photo-documentation to share with the audience in the packed auditorium. His firsthand account clarified to me in a very visceral sense the existential threat of nuclear weapons to our collective future. It also provided me with a glimmer of hope that dialogue was possible and that threats could be collaborative addressed.

36 C. Jones. 2009. Gaia Bites Back: Accelerated Warming. *Futures*, 41, 10, 723–730.

37 A checklist for pandemic influenza risk and impact management: building capacity for pandemic response. Geneva: World Health Organization; 2018. Licence: CC BY-NC-SA 3.0 IGO Cataloguing-in-Publication. Retrieved from https://apps.who.int/iris/bitstream/handle/10665/259884/9789241513623-eng.pdf?sequence=1

38 The COVID-19 pandemic of 2020 is discussed more in the Epilogue.

39 C. Nathan. 2004. Antibiotics at the Crossroads. *Nature*, 431, 899–902.

40 J. Thacker and C. Artlett. 2012. The Law of Unintended Consequences and Antibiotics. *Open Journal of Immunology*, 2, 59–64. https://doi.org/10.4236/oji.2012.22007.

41 W. Klietmann and K. Ruoff. 2001. Bioterrorism: Implications for the Clinical Microbiologist. *Clinical Microbiology Reviews*, 14, 2, 364–381. https://doi.org/10.1128/CMR.14.2.364-381.2001

42 A. Regalado. 2016. How the FBI Polices Do-It-Yourself Biology Labs. *MIT Technology Review*, October 20. Retrieved from www.technologyreview.com/s/602643/on-patrol-with-americas-top-bioterror-cop/

43 P. Torres. 2018. Agential Risks and Information Hazards: An Unavoidable But Dangerous Topic? *Futures*, 95, 86–97.

44 B. Plummer. 2017. Humans Are Bringing about the Sixth Mass Extinction of Life on Earth, Scientists Warn. *The Independent, Independent Digital News and Media*, May 31. Retrieved from www.independent.co.uk/environment/mass-extinction-humans-causing-earth-deaths-end-times-warning-a7765856.html.

45 Millennium Ecosystem Assessment. 2005. *Ecosystems and Human Well-being: Biodiversity Synthesis.* World Resources Institute, Washington, DC. Retrieved from www.millenniumassessment.org/documents/document.354.aspx.pdf

46 Biodiversity International. 2017. *Mainstreaming Agrobiodiversity in Sustainable Food Systems: Scientific Foundations for an Agrobiodiversity Index.* Biodiversity International, Rome (Italy), 180 p. ISBN: 978-92-9255-070-7

47 D. Denkenberger. 2015. Feeding Everyone: Solving the Food Crisis in Event of Global Catastrophes That Kill Crops or Obscure the Sun. *Futures*, 72, 57–68.

48 Food and Agricultural Organization. 2020. Background | FAO's Global Action on Pollination Services for Sustainable Agriculture | Food and Agriculture Organization of the United Nations. Retrieved from www.fao.org/pollination/background/en

49 J. Diamond. 2005. *Collapse: How Societies Choose to Fail or Succeed.* Penguin Books, New York.

50 www.stockholmresilience.org/research/planetary-boundaries/planetary-boundaries/about-the-research/the-nine-planetary-boundaries.html

51 Baum and Handoh propose a framework that integrates the planetary boundary and global catastrophic risk paradigms. S. Baum and I. Handoh. 2014. Integrating the Planetary Boundaries and Global Catastrophic Risk Paradigms. *Ecological Economics*, 107, 13–21.

52 Almost a Quarter of all Disease Caused by Environmental Exposure. 2011, February 14. Retrieved from www.who.int/mediacentre/news/releases/2006/pr32/en/

53 J. Blomberg, L. Priskorn, T. Jensen, A. Juul, and N. Skakkebaek. 2015. Temporal Trends in Fertility Rates: A Nationwide Registry Based Study from 1901 to 2014. *PLOS ONE* 10, 12, e0143722. https://doi.org/10.1371/journal.pone.0143722

54 T. Lopes, T. Chermack, D. Demers, M. Kari, B. Kasshanna, and T. Payne. 2009. Human Extinction Scenario Frameworks. *Futures*, 41, 10, 731–737.

55 A. Vidyasagar. 2018. What Is CRISPR? *LiveScience*, Purch, April 21. Retrieved from www.livescience.com/58790-crispr-explained.html; J. Doudna and S. Sternberg. 2017. *A Crack in Creation: Gene Editing and the Unthinkable Power to Control Evolution*. Houghton Mifflin Harcourt, New York.

56 Dysgenics. 2019. *Wikipedia*. Wikimedia Foundation, December 18. Retrieved from en.wikipedia.org/wiki/Dysgenics.

57 Total Fertility Rate Population. (2020, February 17). Retrieved March 11, 2020, from http://worldpopulationreview.com/countries/total-fertility-rate/

58 Ibid.

59 Government policies could also directly limit reproduction. China's One-Child policy is an example of state-limited human reproduction. Admittedly, it was implemented to contain China's unsustainable population growth, not to lead to human extinction. However, such programs could be implemented in the future by a host of countries that could lead to human extinction.

60 W. Bainbridge. 2009. Demographic Collapse. *Futures*, 41, 738–745.

61 Transhumanism. 2020. *Wikipedia*. Wikimedia Foundation, March 7. Retrieved from en.wikipedia.org/wiki/Transhumanism.

62 N. Bostrom. 2003. Ethical Issues in Advanced Artificial Intelligence. In I. Smit et al. (Eds.), *Cognitive, Emotive and Ethical Aspects of Decision Making in Humans and in Artificial Intelligence*, vol. 2. Int. Institute of Advanced Studies in Systems Research and Cybernetics, 12–17. Retrieved from https://nickbostrom.com/ethics/ai.html

63 I have argued that we would not be human and that the unanticipated unintended consequence of uploading could turn out to be pathological personality disorders. B. E. Tonn. 2012. Will Psychological Disorders Afflict Uploaded Personalities? *World Future Review*, 3, 4, 25–34.

64 K. Kelly. 2011. *What Technology Wants*. Penguin Books, New York.

65 N. Bostrom. 2014. *Superintelligence: Paths, Dangers and Strategies*. Oxford University Press, Oxford; V. Vinge. 1993. The Technological Singularity, VISION-21 Symposium sponsored by NASA Lewis Research Center and the Ohio Aerospace Institute, March 30–31. Retrieved January 28, 2006, from www.ugcs.caltech.edu/phoenix/Lit/vinge-sing.html

66 R. Kurzweil. 2006. *The Singularity Is Near: When Humans Transcend Biology*. Penguin Books, New York.

67 B. Joy. 2000. Why the Future Doesn't Need Us. *Wired*, April, Issue 8.04.

68 A. Dar, A. De Rujula, and U. Heinz. 1999. Will Relativistic Heavy Ion Colliders Destroy Our Planet? *Physical Letters B*, 470, 142–148; W. Busza, R. Faffe, J. Sandweiss, and F. Wilczek. 2000. Review of Speculative 'Disaster Scenarios' at RHIC. Retrieved from www.arxiv.org/abs/hep-ph/9910333v1

69 www.heritagedaily.com/2019/07/10-of-the-largest-super-volcanoes/124137

70 History.com Editors. 2015. Ice Age. *History.com*, A&E Television Networks, March 11. Retrieved from www.history.com/topics/pre-history/ice-age.

71 www.giss.nasa.gov/research/features/201508_slushball/
72 Anoxic Event. 2015. *Wikipedia*, Wikimedia Foundation, June 26. Retrieved from simple.wikipedia.org/wiki/Anoxic_event.
73 P. Ward. 2004. *Gorgon: Paleontology, Obsession, and the Greatest Catastrophe in Earth's History*. Viking, New York.
74 L. W. Alvarez. 1983. Experimental Evidence That An Asteroid Impact Led to the Extinction of Many Species 65 Million Years Ago. *Proceedings of the National Academy of Sciences of the United States of America*, 80, 2, 627–642. https://doi.org/10.1073/pnas.80.2.627
75 M. Weisberger. 2018. Global Warming vs. Solar Cooling: The Showdown Begins in 2020. *LiveScience*, Purch. Retrieved from www.livescience.com/61716-sun-cooling-global-warming.html
76 Gamma-ray Burst. 2020. Retrieved from https://en.wikipedia.org/wiki/Gamma-ray_burst
77 A. Klesman. 2018. A Nearby Supernova May Have Caused a Mass Extinction 2.6 Million Years Ago. *Discover Magazine*. Retrieved from www.discovermagazine.com/the-sciences/a-nearby-supernova-may-have-caused-a-mass-extinction-26-million-years-ago
78 T. Fish. 2019. Black Hole Shock: A Rogue Black Hole as Big as Jupiter Is Rampaging Through the Milky Way, July 25. Retrieved from www.express.co.uk/news/science/1156927/black-hole-big-as-jupiter-milky-way-space-news
79 C. Lui. 2015. *The Dark Forest. Tr. Joel Martinsen*. Tor Books, New York. ISB-10:076537708X; ISBN-13 978-0763577081
80 K. Mack. 2015. Vacuum Decay: The Ultimate Catastrophe | Cosmos. *Cosmosmagazine.com*. Retrieved from https://cosmosmagazine.com/physics/vacuum-decay-ultimate-catastrophe.
81 E. Gough. 2013. This Is What It'll Look Like When The Milky Way and Andromeda Galaxies Collide Billions Of Years From Now – Universe Today. *Universe Today*. Retrieved from www.universetoday.com/141750/this-is-what-itll-look-like-when-the-milky-way-and-andromeda-galaxies-collide-billions-of-years-from-now/.
82 K. Mack. 2020. *The End of Everything (Astrophysically Speaking)*. Scribner, New York.
83 Ibid.
84 Moore's Law is the observation that the number of transistors in a dense integrated circuit doubles every two years, thereby also increasing computational power as well. https://en.wikipedia.org/wiki/Moore%27s_law

4 Our fate is in our hands

Introduction

One positive lesson that can be taken from the previous chapter is that it is possible for humanity to anticipate any number and type of extinction-level events. Many of these events are within humanity's ability to prevent. Even the unpreventable events can be survivable given our will and ability to adapt. This chapter takes a first crack at building an analytical framework to deal with global catastrophic risks and extinction-level events that account for human behavior. The first section presents a framework for addressing risks that is designed around the concept of resilience. The second section addresses how unfortunate chains of events (CEs) can lead to large losses of life and even our own extinction. Singular Chains of Events Scenarios (SCES) are introduced as a tool to better understand CEs.[1]

It is not enough to work to short-circuit CE. Humanity also needs to be cognizant of potentially destructive unanticipated unintended consequences (UUCs) that could result from our own behavior and, especially, from the rush to market powerful new technologies. Several of these have already been mentioned, including how a melding of technologies could lead to an erosion of what it means to be human if not the disappearance of humans altogether. The goal of the third section is to help us anticipate what seems not to be anticipatable.

The fourth section addresses an even more nebulous issue, unknown unknowns (UKUKs). These are events about which we currently have no knowledge of their current or potential existence and, of course, have no knowledge of their consequences. However, lack of knowledge is not an excuse for us to downplay or totally ignore programs to identify and develop insights about UKUKs.

This chapter concludes a proposal to develop the World Earth Monitoring System (WEMS). This system would integrate inputs that comprehensively describe and measure both the physical and human environments. Among many purposes of WEMS will be to help humanity detect early signals that could indicate the emergence of UUCs and UKUKs.

Framework for addressing risks

Resilient systems and societies are prepared to deal with and rebound from adversity.[2] Figure 4.1 graphically illustrates the concept of resilience. Resilience is defined as response to an initiating event, in this case an extinction-level event, with respect to one or more essential metrics. In this example, the metric is the human population on the earth. This model can be generalized to address non-extinction-level events, such as droughts, wildfires, job loss, and even the loss of a significant other.

Four paths through time are presented in response to a hypothetical extinction-level event. In the perfect world, as depicted by Path A, the extinction-level event has no impact, as there is no loss of life. In Path B, humanity is able to resist the event's impacts for a period of time before the population begins to decline. The decline stabilizes for another period of time and then due to timely and effective adaptation the population fully rebounds. Path C illustrates a generic chain of events (CE) scenario where there is a cascading failure of adaptations, ultimately leading to human extinction. Path D exemplifies a scenario where humanity was totally defenseless (e.g., from a gamma ray burst) and quickly went extinct before being able to adapt in any way.

It can be considered best practice for humanity to engage in these four activities in order to most effectively address extinction-level events:

(1) *Anticipation* – The first step is to anticipate events that could have a major impact on humanity or whatever it the context of concern (e.g., your organization, your country).

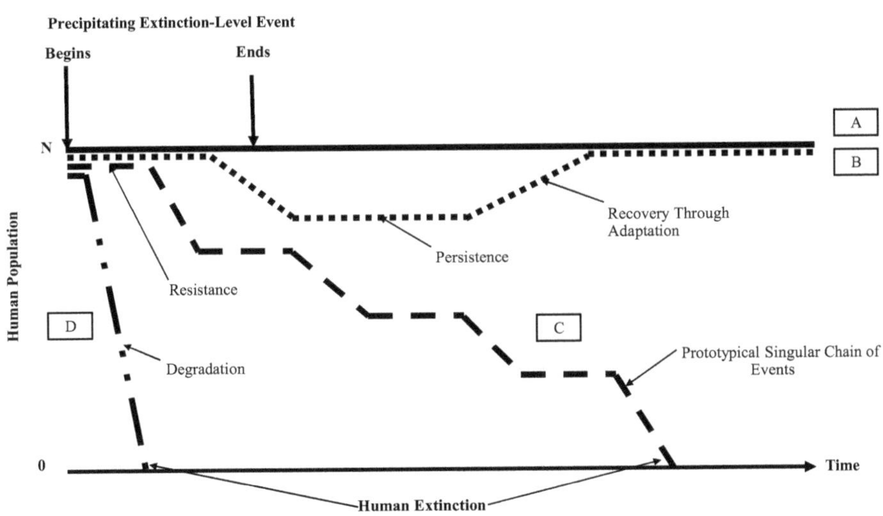

Figure 4.1 Resilience and Human Extinction

(2) *Planning* – If the anticipated impact of an event is substantial, then resources should be devoted to planning courses of action to deal with the event. Plans can be made to mitigate the negative impacts of the event or adapting to its occurrence.
(3) *Mitigation efforts* – The plans should then be implemented. Some components of the plans may focus on preventing events and/or mitigating the magnitude of negative consequences that could be caused by the events. Referring to Path B in Figure 4.1, the latter actions, if successful, will increase resistance and also decrease degradation.
(4) *Adaptation* – If mitigation efforts appear to be inadequate to protect harm from the precipitating event, then adaptation actions need to be implemented. Successful adaptation actions can also increase resistance and decrease degradation as well as reduce the time period of persistent harm and increase recovery.

Climate change provides an excellent case study to assess this framework. It is clear that humanity is actively engaged in anticipating the negative impacts of climate change. One could argue that the Intergovernmental Panel on Climate Change (IPCC) is the key coordination node for these efforts. Plans have been and are being developed across the globe to mitigate the emissions of GHG and to adapt to climate change. For example, actions are being implemented to reduce GHG emissions, such as by substituting renewable energy resources for fossil fuels. Adaptation actions are also being implemented. These include building dikes around low-lying urban areas and hardening electric power systems. Unfortunately, these efforts may not be anywhere near enough at this point in time to prevent major harm to humanity from climate change.

Table 4.1 presents my subjective judgments about how well humanity is engaging in these four activities with respect to each of the extinction. Also

Table 4.1 Assessment of How Well Humanity is Dealing with Extinction-Level Events

Source of Existential Risk/Evaluation Criteria	*Societal Status of Anticipation*	*Societal Status of Planning*	*Societal Status of Mitigation Efforts*	*Current Ability to Adapt*	*Current Estimate of Survivability*
I. Anthropogenic	P	C	C	F	C
II. Coupled Human-Env. SYS	C	C	F	F	C
III. Human Reproduction	F	F	F	F	C
IV. Risk to Humanness	F	F	F	F	C
V. Advanced Technology	F	F	F	F	C
VI. Natural Terrestrial	F	F	F	F	F
VII. Solar System	F	F	F	F	F
VIII. Extraterrestrial	F	F	F	F	F
IX. Universe Scale	F	F	F	F	F

Legend: P (pass) – things are fine; C (concerned) – we should be concerned; and F (fail) – we are completely failing.

provided is an estimate of how survivable each category is in light of these judgments. The entry for each cell is assessed on a three-element scale: P (pass) – things are fine; C (concerned) – we should be concerned; and F (fail) – we are completely failing.

As expected, most of the scores are fail. Overall, humanity is not engaged in rigorous anticipation programs with respect to human extinction. We have not implemented planning processes to deal with the extinction-level risks nor have we invested in adaptation to allow us to better survive extinction-level events. Turning these scores from fail to pass may take hundreds if not thousands of years.

Singular chains of events scenarios

Scenarios are used to study futures. A typical scenario depicts a potential future world five to 20 years into the future. The more compelling scenarios have provocative names and are communicated in story form though a narrator, as a news article, or even as a diary entry. Sets of scenarios are often developed to help organizations of every conceivable type, from businesses to governments to non-profits, build forward-looking strategies. Scenarios within a set are designed to be quite distinct so as to capture the space of futures that the organization might have to confront. Distinctions between scenarios are driven by different assumptions about the influences of important societal, technological, and environmental trends. Strong organizational strategies are those that seem to be positively applicable across most if not all of the scenarios.[3]

Global scenarios have also been developed to aid policy makers. For example, the IPCC developed emission scenarios to help guide climate change research.[4] The International Energy Agency produces the World Energy Outlook that contains global energy scenarios out several decades.[5] The Millennium Project is known for developing comprehensive sets of global scenarios on various topics, such as the future of work and technology, also several decades out.[6]

Scenarios can also be developed to help us understand the risks of human extinction. These scenarios are different from typical scenarios in that they all end up at the same place, human extinction. Also, because the extinction stories may encompass exotic risks and very long timescales, human extinction scenarios may at times take on the characteristics of science fiction. This can certainly be said of the extinction scenario presented at the end of this chapter.

Another characteristic that extinction scenarios can share is their focus on failures of humanity to deal with CEs that could start with the occurrence of extinction-level events or encompass such events at any step along the way.[7] These scenarios assume that, following Murphy's Law, almost everything that could go wrong does go wrong. These scenarios are referred to herein as *Singular* Chains of Events Scenarios (SCES) because each in its own way is unique and remarkable.

Figure 4.2 presents a second way of visualizing Path C. There is a precipitating extinction-level event or even an event that at first glance does not seem to be catastrophic in and of itself. However, humanity's response to this first event is inadequate in some fashion. Then something else happens. Humanity's ability to respond to the second event is degraded (i.e., we are not as resilient), and we also fail to adequately deal with this situation. Additional events occur, some within our ability to prevent but we do not, some we cannot prevent and are quite unlucky to have them occur with our resilience degraded. Eventually, the CEs result in extinction.

The purpose of SCES is to add important details to generic CEs. Extinction scenarios are an understudied component of futures studies, though several have been written and published. The apocalypse has been a staple of human thought for millennia, but fears of end times were rarely depicted in explicit stories about the how the last human becomes extinct. I believe that the very first human extinction scenario was actually the book, *The Last Man*, published in 1826 by the author of *Frankenstein*, Mary Wollstonecraft Shelly (MWS).[8] The book chronicles the life and times of Lionel Verney who paid witness to the death of all humans, except himself, from a mysterious Plague that was exacerbated by adverse climatic conditions. MWS shows us how everything collapses and disappears; civil society, culture, the arts, all of mankind's past accomplishments. The novel is very apt for any generation's consideration of existential risk.[9]

MSW's novel was the penultimate contribution to a series of pieces during the early nineteenth century around the theme of the *Last Man*.[10] This time period in Western culture was rife with change, fear, and uncertainties. Many people thought that the apocalypse was imminent. Many were not

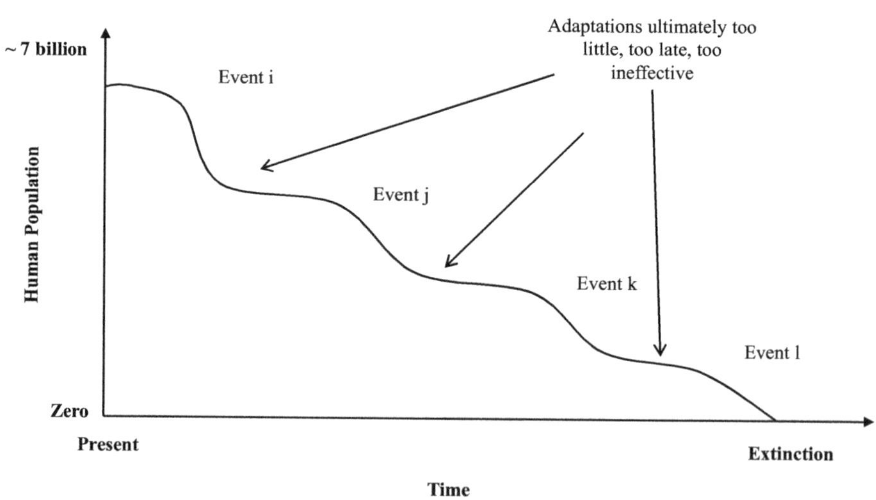

Figure 4.2 Visualizing a Generic Chain of Events Leading to Extinction

so worried about this, as many also believed that the apocalypse would be followed by a millennium of rebirth and progress. MWS's novel broke with convention by ending the book without the hopeful millennium, as the Last Man sailed off to Africa to see if anyone else was still alive.

In a special issue of *Futures*, Morgan,[11] Lopes et al.,[12] Carpenter and Bishop,[13] Bainbridge,[14] Jones,[15] Goux-Baudiment,[16] and Tonn and MacGregor[17] all presented human extinction scenarios. In every case, the human extinction scenario ends in failure; humans are unable to prevent their own demise. Humans are faced with a series of threats that in the end we are unable to overcome. For various reasons, we may not take these threats seriously, may act on them too late, may implement poorly thought-out responses, and/or simply may not know how to properly react to the threats. Our adaptations generally are too little and too late.

For a human extinction scenario to be of value to policy makers, it must possess verisimilitude. That is, the threats included in the scenario need to be plausible. The scenario in its entirety may not be very probable, which is fine because much can be learned from improbable scenarios. The driving forces behind the threats need to be well explained. Human responses to the threats need to be probed. The path of events and adaptations must show how today's human population that exceeds six billion people decreases over time until the last human in the universe breathes their last breath. Most importantly, the scenario must be able to suggest what current generations could do now to help reduce the probability of the particular path from occurring.

Figure 4.3 presents an SCES that has these components.[18] The first phase encompasses a set of economic and social events that lead to a rapid and substantial decline in human population. There is a key trigger event, the destruction of an irreplaceable global economic resource that sets off a domino effect of other events that encompass the collapse of political and economic institutions worldwide. The ensuing chaos results in the breakdown of systems that maintain human health and safety. The breakdown of these systems, along with global environmental change, leads to massive levels of human death due to pandemics and starvation. The actual scenario specifies which key global economic resource was destroyed – oil – why political institutions collapsed, and how a new flu pandemic came about. However, we believe that these specific details in this block of the scenario could easily be changed (e.g., a different key trigger could lead to political and economic chaos, a different set of health problems could have devastated the population, etc.) without altering the fundamental causal structure of the scenario.

The second phase posits the decline of human civilization and its technological base. Here it is assumed that for various reasons humans are unable to bounce back. We are unable to regain our technological prowess. Our knowledge base becomes lost to time. Great civilizations crumble and the future begins to resemble a more feudal and tribal past. People

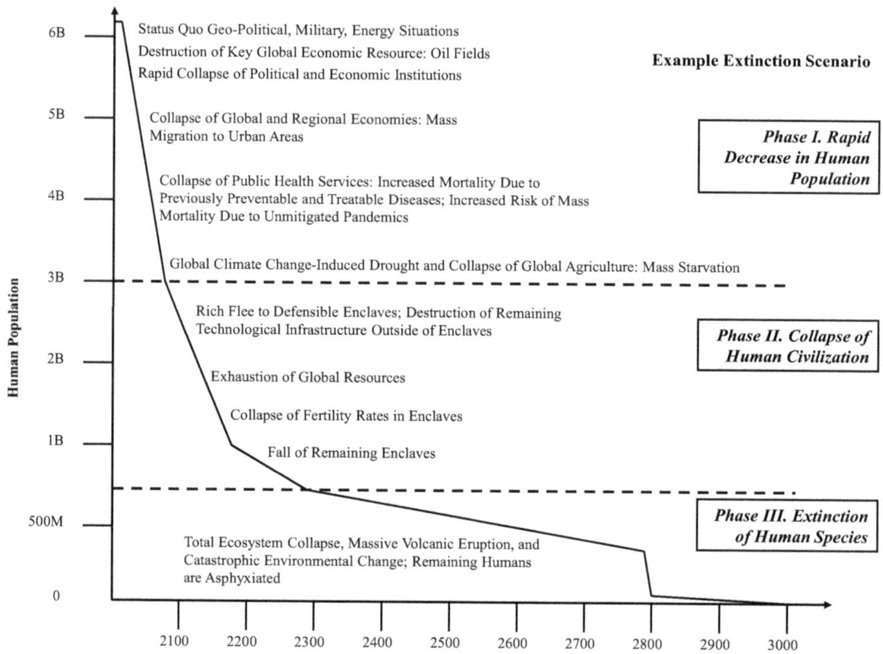

Figure 4.3 A Detailed Singular Chain of Events Scenario

congregate into groups and enclaves and fight each other to protect their own resources and future. No one emerges from the chaos to help humans cooperate toward a better future. Population continues to decline because our technological and institutional capabilities have declined. Ecological catastrophes and other disasters also continue to take a toll on the human population. The number of humans on the earth begins to move toward the number on the earth before the Agricultural Revolution.

The third phase is the end-game pathway. This phase explains how the relatively few humans still surviving meet their end. In our scenario, a random event, a massive volcanic eruption, triggers severe climate change, which leaves only a few thousand humans on the earth. Then, the remaining humans are unable to adapt successfully to further global environmental change.

There are ten specific entries in the three phases of Figure 4.3. If one were to estimate the likelihood of the precipitating event at, say 1 in a 1000 and estimate the likelihood of each succeeding event conditional upon the previous event somewhere between 1 and 1000 and 1 in 10, then the likelihood of the entire chain could be estimated to be $10–^{20}$. So, this path to extinction appears to be quite unlikely. How serious humanity should take this observation depends upon how high a risk of our own extinction we are willing to accept. The last section of Chapter 6 develops an approach to estimating an ethical threshold of human extinction based on three perpetual obligations presented earlier in the chapter. For now, the important point is that this approach suggests that the ethical threshold is also 10^{-20}. In other words, if there are only two such paths to extinction for every 10^{20} paths humanity could take into the future, then humanity would be in breach of the 10^{-20} threshold.

Figure 4.4 illustrates this phenomenon using a cone of possible worlds over time. Now, there is only one world, the one we are living in. But, projecting into the future, the number of possible futures increases exponentially, based upon decisions we make, new technologies created, extinction events occurring, etc. The dashed paths into the future are good; they do not result in extinction. The dotted ones are bad; they do lead to extinction. Even though the number of dashed paths could be enormous, a relatively small number of dotted paths could suggest we are not meeting our obligation to keep the risk of human extinction below the ethical threshold.

Let's refer back to Figure 4.3. The initiation and progression of each SCES need to be based on plausible events and drivers. Have we witnessed the events and drivers in the past and/or can we anticipate the events using current scientific knowledge? If so, those events and drivers can be used in a human extinction scenario. Does the scenario allow for human adaptation? It should. In this case there are numerous policies that could short-circuit this SCES, from moving to distributed renewable energy resources to substantially increasing the resilience of crops. Substantially more effort is needed to deal with the impacts of the super-volcano eruption. Of course, the key characteristic of a human extinction scenario is that all adaptations

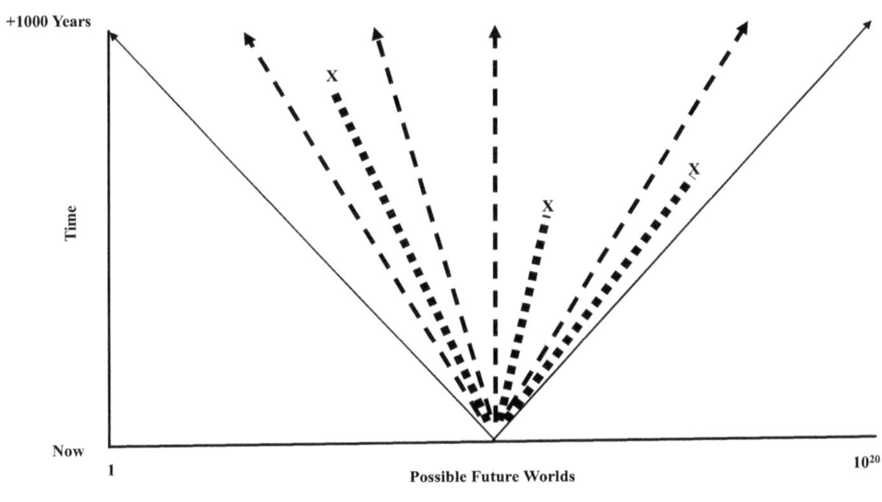

Figure 4.4 Cone of Possible Worlds

Legend: Survival path (Dashed) and Extinction path (Dotted).

eventually fail, as should be clearly explained in the scenario itself. This scenario is composed of plausible components even though its overall probability is probably quite small.

Unintended consequences[19]

Up to this point, we have focused almost exclusively on specifically identifiable existential risks such as climate change and pandemics. It has been suggested that the probability of existential risk is fluid, in large part dependent upon our own abilities for anticipation, planning, and successful adaptation with respect to identifiable existential risks and cascades of additional events and decisions. This section expands this previous discussion by adding another component: unintended consequences.

Merton published the first substantive and thoughtful piece on unintended consequences of purposive actions.[20] A purposive action (hereinafter 'action') can be the introduction of a new technology, whether based on new scientific discoveries or a combination of existing technologies,[21] where technology is broadly defined to be a human-manufactured artifact or system or a human intervention into a system (e.g., ecosystem, biological, energy), usually directly or indirectly related to a technological-type policy decision. Any science or technology can be introduced by anyone, anytime, anywhere. An action can also be a policy decision or major program. For example, policy responses captured in Figures 4.1 (Path C) and 4.2 that had unintended consequences could be considered actions in this framework.

The consequences of purposive actions are some combination of three types of consequences:

(1) Anticipated-intended consequence(s) (AIC) – Represent the intended purposes of the action.
(2) Anticipated-unintended consequence(s) (AUC) – Represent anticipated side effects, positive or negative, that are consequences of the action.
(3) Unanticipated-unintended consequence(s) (UUC) – Represent consequences, positive or negative, that historical or future records suggest were not anticipated or even remotely imagined by the initiators of the action.

We are concerned with improving our ability to imagine and therefore anticipate more rather than fewer unintended consequences of actions that might create pathways to extinction or global catastrophes. We are also concerned with developing approaches to dealing with as yet unimaginable unintended consequences (i.e., anticipating there will be UUCs).

The unintended consequences of an action or event might relate to a lack of knowledge (i.e., due to the stochastic nature of social behavior or due to the lack of knowledge related to the cost of producing such knowledge);

errors in judgment; rush to immediate action; biases or values that overcome other considerations; or self-defeating prophecies.[22] The action might be the accumulation,[23] or the interdependence, or the spillover of actions.[24] For example, as more and more people take the action to visit a mountain retreat for solitude, the unintended consequence of the quest for solitude is the loss of that solitude.[25] Connecting the action or an event with the UUCs is challenging as some amount of time often intervenes[26] and neither the magnitude nor the level of uncertainty provides enough information to differentiate risks (Mazri 2017).[27]

The human inability to adequately deal with the unexpected "embarrasses our capacity and willingness to predict" and "insults our competency to control."[28] This also is why smaller actions to mitigate and adapt to consequences, with presumably smaller unintended consequences and more opportunity to mitigate unintended consequences, are typically recommended. For example, many argue that the more global the solution to climate change is, such as massive geoengineering projects, the more prone they are to undesirable UUCs.[29] Smaller approaches, such as removing carbon from the atmosphere, make it easier to identify UUCs.

Science and technology artifacts that spin off anticipated and UUCs are common. Photographic technology, as intended, successfully captures and stores images. Elevators successfully transport people, pets, and objects up and down through tall buildings. Bows successfully propel arrows to the intended targets. In contrast, splitting the atom to produce the atomic bomb had the unintended but anticipated consequence of creating the nuclear power industry. UUCs with respect to this industry are prodigious amounts of dangerous nuclear wastes and the resistance of states and communities in the United States to hosting permanent nuclear waste repositories and other nuclear facilities, such as breeder reactors.[30] Mao's environmental policy to eradicate sparrows was designed to reduce agricultural losses (AIC), yet it had devastating impacts on the ecosystem in general and on agriculture yields in particular as farmers learned that sparrows were critical to controlling other pests (UUCs).[31]

The unintended consequences framework explicitly incorporates existential risks of human extinction because science and technology are the underlying causes for a large number of these extinction events (e.g., categories I–V). Moreover, converging science and technology also will contribute to our existential risk burden. Applications of existing or new technologies to ameliorate existential risks are likely to result in additional AUCs and UUCs, including additional existential risks of human extinction.

In common with CEs, scenarios are the primary analytic method proposed to help us better understand UUCs. Here are four scenario approaches that are especially appropriate for generating this content:

(1) Evolution-Over-Time Scenarios;
(2) Market-Saturation Scenarios;

(3) Interventions in Tightly Coupled Systems Scenarios; and
(4) Existential-Risk-of-Human-Extinction Scenarios.

The next several subsections apply these four scenario approaches to understanding UUCs that surfaced from current technologies and to emerging and prospective technologies.

Evolution-over-time scenarios

Evolution-over-time scenarios depict the ways in which a scientific or technological action could play out over time. A useful image to guide the development of these scenarios is the iconic market-penetration S-curve: As the technology infuses itself into the world, what might be the unintended consequences? These scenarios focus on describing three things:

(1) Inputs needed by the technology[32];
(2) How the technology is used and/or operates in the world; and
(3) Outputs spun off by the technology (e.g., pollutants).

The goal of this exercise is to identify potential AUCs and generic UUCs that could emerge during the transitional growth stage of the technology. Plans can be made to deal with the AUCs, and processes could be implemented to detect weak signals early on to address UUCs.[33] Also, the more AUCs imagined, the more that researchers and policy makers should be concerned about UUCs. Automobile technologies are examples of existing technologies that can provide insights into the development of these evolution-over-time scenarios.[34]

The inventors and early proponents of the automobile envisioned a technology to move people and goods more quickly than walking or riding horses and with more flexibility than trains and streetcars. Certainly, the AICs of this technology are borne out: hundreds of millions of automobiles perform this function the world over every hour of every day. An AUC of this technology is that the automobile industry requires capricious amounts of inputs, from raw materials to produce steel, aluminum, and glass to refined petroleum products to power internal combustion engines. It is the demand for the latter that has spiked worries about fast-approaching peak oil thresholds[35] and has led to UUCs including oil embargoes and military interventions to ensure access to oil from the Middle East.

Automobiles also demand extensive infrastructure, the sheer scale of which can be characterized as an UUC. Production of the concrete needed to build roads and highways emits prodigious amounts of carbon into the atmosphere, another UUC. Automobiles, trucks, and buses contribute substantially to the buildup of GHG in the atmosphere, which was identified as an extinction-level event in Chapter 3. Another UUC of the automobile is sprawl, which locks in the need for the automobile and the resulting GHG emissions. An evolution-over-time scenario could have envisioned trends in human relocation (sprawl) and the resulting carbon emissions from autos and the production of concrete. Policies could have been put in place during the 1960s, for instance, to incentivize urban *versus* suburban living to proactively deal with these problems.

Autonomous vehicles (AVs) are examples of emerging technologies that can provide insights into the development of these evolution-over-time scenarios. The most common image of AVs are self-driving cars, although there also are AVs that fly, float, and submerge. Relative to the self-driving car: what might be learned from imagining how AVs could evolve over time and then produce unintended and unanticipated consequences?

First, let us assume that the major intended consequences of AVs are to improve driving safety and lessen the drudgery of driving. If these goals are met, then, for example, Rubin argues that AUC of AVs include additional energy use and emissions of GHG; decreases in urban density; and increases in congestion because people will drive more, be more willing to have longer commutes, and even to sit in traffic, respectively.[36] It is also widely anticipated that AVs will reduce employment opportunities for truck, taxi, ride-sharing, and other drivers.

As the technology matures and people become more habituated to allowing their vehicles to drive, the form of vehicles could change. Why should passengers sit in uncomfortable seats facing the front of the vehicle when they could sit on chairs at desks doing work or watching television? Vehicles could begin to resemble recreational vehicles, with beds, appliances, and exercise equipment. Given the popularity of current ride-sharing options, over time, vehicle ownership could give way to massive levels of ride-sharing and would-be passengers could hail various types of miniature autonomous recreational vehicles depending on their needs in real time (e.g., a coffee dispensing AV, an office AV, a gym AV). The proliferation of miniature recreational vehicles may then require changes to the physical infrastructure of the transportation system, from parking garages to fueling stations to driveways, all AUCs.

The systems supporting AVs will evolve over time, not only to be able to perform more reliably but also to deal with ethical decision-making in real time. As depicted in the scenario presented in the previous chapter, *Death by Autonomous Vehicles*, the evolution of these capabilities could potentially pose an existential threat to humans if it creates super artificial intelligences; humans are unable to detect or understand computer 'language' interactions; and/or AVs figure out that the best way to assure their existence is through the coordinated subjugation if not outright extinction of humans.[37] AV technology illustrates how AUCs of small-to-existential threat-level magnitudes can be generated at every step in the evolution of new technologies: fewer cars leads to fewer driving jobs, which leads to increased need for regulation of the vehicles.

Market-saturation scenarios for existing technologies

Market-saturation scenarios depict the ways in which a scientific or technological action could play out if the science or technology completely saturates a market. The thought is UUCs can be best understood only when the market is completely saturated. Here are several examples to help explain this concept:

- *Urban water works*: The AIC was safe drinking water and disposal of waste, with huge advances in public health and lives saved. The UUC was a huge increase in the global human population resulting from longer lifespans.
- *Birth control*: The AIC was giving humans choices about reproduction. The UUCs were that birth levels in many developed countries are now below replacement rates (OECD [Organisation for Economic Co-operation and Development] 2017). Moreover, although oral contraceptive pills are a tiny cause of estrogenic chemicals in waterways and drinking water, they are joining estrogenic chemicals from agriculture and industry and those consequences are still emerging.[38]

Television is another example of an existing technology that can provide insights into the development of these market-saturation scenarios. The inventors of television could not have envisioned the degree to which this technology came to saturate the market: in 2010, 98% of the homes in the United States owned at least one television.[39] Optimists viewed this technology as the perfect vehicle for citizen education and enlightenment, the AIC, yet it did not take long for critics to describe commercial television as a "vast wasteland,"[40] an UUC.

Television demands inputs from its users, several hours per day, every day, at every time of day. The UUC is that the more time that individuals spend watching television, the fewer hours that they spend doing other things. In his seminal work, *Bowling Alone,* Putman described the steady decline of social capital in the United States and blamed the television in part.[41] People began staying at home to watch television instead of joining bowling leagues or otherwise volunteering their time through social groups. Personally, having grown up with television in my home, my default behavior is to go home after work and watch a couple of hours of television rather than venture out to interact with people in ways to build social capital. Additionally, instead of providing education and enlightenment – or being simply innocuous – television was tearing at the social fabric of society. The advent of new forms of social media is likely exacerbating this situation.

Drones are examples of emerging technologies that can provide insights into the development of these market-saturation scenarios. Drones are unmanned aircraft of various sizes that are remotely controlled or can fly autonomously. The most iconic drones are those used in military situations, by hobbyists, or proposed by Amazon to deliver products. In the drone market-saturation scenario, the AIC of drones is to quickly and efficiently move things, such as food and books. To begin to describe market saturation, let's imagine that households have multiple drones that fetch goods from retailors, friends, and relatives in addition to retailers of every type having their own fleets of drones. Suppose that Federal Express, the United Parcel Service, and even the United States Postal Service all deliver letters and packages via drones. Groceries are delivered by drones; a few goods at a time but continuously throughout the week. Local organic farmers use drones to deliver just-picked fruits and vegetables, even single freshly laid eggs. Configurations of drones are assembled to move ever-larger artifacts, such as tables. Governments also use drones to monitor air quality, traffic conditions, and public safety, among other uses. A metropolitan area with a million inhabitants, which has become accustomed to 800,000 or more ground-bound vehicles, now must deal with a million or more airborne drones.

The potential AUCs of having a major urban area inundated with drones might include:

- Crashes of drones in mid-air resulting in the loss of drones and their goods;
- New laws to deal with liability issues arising from drone accidents;
- Safety issues on the ground that affect pedestrians, vehicles, buildings hit by moving drones or drones dropping out of the sky or the packages being dropped by drones;
- Bird populations hit by drones;
- Noise issues with drones flying overhead continuously;
- New policies to deal with stranded drones that may have landed because of mechanical failures or drained batteries;
- Aesthetic issues with once-clear skies that are now blighted by flocks of drones;
- Drone piracy, with perpetrators capturing drones in nets to pirate their treasures, for example, or ambushing drone landings; and
- Alterations to homes and businesses to facilitate the egress and ingress of drones and goods. Required changes in zoning laws, building codes, and homeowners' agreements may be needed to allow these alterations.

Although not every AUC listed here would occur in a market-saturation scenario, the list of potential unintended consequences is bound to be extensive given that so many can already be hypothesized at so early a point in their development.

Interventions in tightly coupled systems

In tightly coupled systems scenarios, the event represents the change in a key node in a tightly coupled system or the addition of a new node that immediately establishes connections to key nodes. A change in a key node could represent the elimination of the node or a radical change in that node's capabilities. The goal of this exercise is to identify potential AUCs and UUCs that could emerge at the changes in key nodes or the additions or new nodes, usually in relation to existing nodes or system characteristics.

The misguided policy to reduce sparrow populations in China has already been introduced as an example. Living in the Southeastern United States, I can personally attest that the introduction of kudzu has had the UUC of becoming a truly annoying invasive species. I am also personally worried about pesticides and herbicides ending up in our water supplies[42] that could become endocrine disruptors. Adding to this list are the not completely understood impacts of other pollutants in our drinking water, which include steroids, caffeine, anti-depressants, and birth control substances.

For a deeper look into the UUCs of actions on tightly coupled systems, let's consider ethanol. In the United States and other countries, corn is grown to produce ethanol, which can be used to supplement or replace gasoline extracted from petroleum resources to power automobiles. In fact, the United States' Renewable Fuel Standard defines the ways in which ethanol must be mixed into the gasoline supply.[43] The AICs of the law were to: decrease United States' reliance on imported foreign oil; reduce emissions of GHG; and increase benefits to agricultural states and farmers. Only the latter AIC has actually emerged, though, as ethanol has little effect on oil imports and rigorous scientific analyses suggest that the production of corn for ethanol actually produces more GHG than are avoided.

The UUCs have been severe. At the extreme, redistributing corn production from food to fuel led to food shortages.[44] Additional acreage devoted to corn production resulted in increased habitat loss,[45] and producing ethanol can incentivize use of forests and wilderness as fuel source.[46]

Genetically altered mosquitoes, using CRISPR technology for instance, are examples of emerging technologies that can provide insights into the development of these interventions in tightly coupled systems scenarios. Scientists are designing mosquitoes to reduce mosquito populations by reducing the transmission of diseases. Google/Alphabet recently announced a project to release altered male mosquitoes that are infected with a bacterium that, while harmless to humans, creates non-hatching dead eggs when they mate with wild females – hopefully cutting the mosquito population and the transmission of the diseases they carry, the AICs.[47] Given the negative unintended consequences described earlier that arise from such interventions into tightly coupled ecosystems, it is likely that major decreases in the number and function of mosquitoes could have serious AUCs, such as increases in their former host populations. For instance, perhaps bird populations would increase if West Nile Virus (carried by mosquitoes) is less prevalent. Perhaps, the lack of mosquitoes would lead to a decline in the bats, which feed on them, and other insect populations could quickly rebound with disastrous consequences for human health, agriculture, etc.

For another example, one proposed plan is to use CRISPR to drive genes into Asian carp, a species that has invaded North American ecosystems. Proponents want to drive a gene into female carp that have them produce only female offspring and, theoretically, cause the local extinction of this species before they invade the Great Lakes. The AUC is that these 'infected' carp could make their way back to Asia and decimate populations there. Or, another UUC, it could work but the gene is transferred to other species with similar immediate effects but unknown cascading effects.

Existential-risk-of-human-extinction scenarios for existing technologies

Existential-risk-of-human-extinction scenarios depict the ways in which a scientific or technological action could end in human extinction. The goal of this exercise is to identify potential UUCs that could produce human extinction or could limit options relative to humans and/or our obligations to future generations. One can consider these scenarios as a subset of SCES.

Domestication of animals is an example of an existing technology that can provide insights into the development of these existential-risk-of-human-extinction scenarios. Diamond describes the link between specific conditions in the world and the related inventions and innovations that then led to the dominance of some societies over others.[48] In particular, the domestication of animals and development of agriculture allowed human societies to transition from low-population hunter-gatherer lifestyles to denser urban environments, an AIC. Moreover, pigs, used to living in their own dense communities, brought to these new human societies transmittable pathogens, namely a host of viruses, an UUC. Human societies have been plagued by zoonotic viruses ever since.

Various humans and human populations have developed immunities to various diseases, though humans exposed to new pathogens are particularly at risk (e.g., the Black Plague and the virtual extinction of Native Americans who were exposed to pathogens carried over the seas by European conquerors). Pathogens jump not only from swine but also from birds (avian flu), bats, monkeys (HIV [Human Immunodeficiency Virus]), and other species. Given the dense network of animals and humans across the globe, the opportunities for diseases to interact with each other are growing and, through a small number of international plane flights, viruses can quickly spread across the globe (e.g., Ebola coming to the United States). Viruses created in the lab greatly magnify the catastrophic potential of extinction of humankind from pandemic.

Chapter 3 has already introduced several advanced and emerging technologies whose UUCs could potentially lead to human extinction. These include super AIs and CRISPR. The *Driven to Extinction* scenario at the end of Chapter 3 describes one such path to human extinction associated with the latter. Let's conclude this discussion by focusing on UUCs potentially attributable to geoengineering. To ameliorate the negative consequences of the building-up of GHG in the atmosphere (i.e., global warming that could have catastrophic consequences for humankind and other species on the earth), most of the approaches to geoengineering entail interventions in the tightly coupled system of systems known as the global environment.

Possible approaches include the following:

- Seeding oceans with iron to promote the growth of microorganisms, which is likely to extract ever larger amounts of carbon from the atmosphere and then sequester the carbon in the bottom of the oceans when they die;
- Seeding the upper atmosphere with sulfur particles to reflect the sunlight back into space, which is likely to reduce solar gain on the earth enough to offset the warming potential of additional GHG in the atmosphere; or
- Deploying large structures in space that are likely to be capable of blocking the sunlight from reaching even the farthest reaches of the earth's atmosphere.

Each of these approaches alters the normal functioning of the earth's environment, the impacts of which then would spread throughout the tightly coupled system of systems that compose the global environment and thereby increase the likelihood of both AUCs. For example, promoting populations of microorganisms may end up increasing the food supply for direct predators and then for species that feed on them. Carbon may end up being sequestered not at the bottom of the oceans but in organisms.

An AUC of seeding the upper atmosphere with sulfur particles is that if this process were ever stopped, the ensuing impacts of global warming could be orders of magnitude worse. The reduction of sunlight by solar shades could have AUCs on photosynthesis, which could completely disrupt regional ecosystems and human psychology from the disruption of circadian rhythms and the depressive aspects of incessant darkness. Implementation of any of these approaches has significant implications for international relations, national security, and the global economy.

Unknown unknowns

Reports that say that something hasn't happened are always interesting to me, because as we know, there are known knowns; there are things we know we know. We also know there are known unknowns; that is to say we know there are some things we do not know. But there are also unknown unknowns – the ones we don't know we don't know. And if one looks throughout the history of our country and other free countries, it is the latter category that tend to be the difficult ones. Donald Rumsfeld[49]

Donald Rumsfeld is credited with introducing the concept of *unknown unknowns* (UKUKs) into the modern lexicon in his reply, quoted earlier, to a question about Iraq and weapons of mass destruction.[50] The concept is fairly self-explanatory: it relates to the fact that there are things out there in the world that are completely unknown to us, and we don't even know we don't know about them. Unknown unknowns are important for our discussion here because they represent the final component of our anticipatory framework.

Unknown unknowns are not, by definition here, unanticipated unintended consequences, which could be argued are known unknowns in Rumsfeld framework (i.e., we know there will be UUCs but we just do not yet know what they are). UKUKs are more like the unknowns referenced in Category VIII of our existential risk framework. We truly have no idea what we are looking for nor what the implications are for what it is we do not know. We can think about UKUKs as having these characteristics:

- They could represent new members of our nine existential risk categories and maybe even additional categories;
- They could be sub-extinction-level events that could exacerbate the seriousness of other events in an SCES;
- They could be events not accounted for in SCES that may accelerate and/or make SCES harder to short-circuit.

Within the field of ecology, UKUKs could represent discontinuities or synergisms.[51] The former is represented by sudden phase changes in an ecosystem, such as when a system completely collapses when the available nutrients falls below a critical threshold. More broadly, one could adapt the market saturation and tightly coupled scenario approaches to search for UKUKs, where systems to explore the consequences are assumed to be maximally stressed. Synergisms are essentially represented by the unpredictable behaviors of non-linear systems.

I would also add that UKUKs could also be linked to the emergence of brand-new systems whose existence could not have been anticipated simply by examining constituent parts. In this sense, UKUKs are an epistemological challenge; we cannot know or understand them until they emerge. The archetypal emergent system is consciousness, which could not have been foreseen simply by studying atoms, molecules, or even collections of cells. Other emergent systems include ant and bee colonies, human cities, and even life itself![52]

We humans have to fight our overconfidence in the comprehensiveness and depth of our current knowledge[53] and anticipatory assumptions[54] so that we can purposively address UKUKs. In the parlance of future studies, we need to be constantly vigilant for weak signals that could suggest the emergence of UKUKs.[55] Searching for weak signals is somewhat like searching for extraterrestrial intelligence. The SETI[56] program used computers worldwide to sift through reams of electromagnetic data to identify potential communications from other intelligent life. Back on the earth, UKUKs may only be detected by being able to imagine new types of data and variables and then being able to create means to measure new variables. It very much requires us to pay attention to any anomalies we see in the environment, even small events or signals that spur us to wonder why this or why that!

UKUKs can also be viewed as black swans or wild cards. A wild card is usually understood to be an event that could completely derail a scenario. Typically, wild cards are known unknowns in that we can describe the event but cannot describe its consequences very well. In the UKUK context, the identity of the wild cards is also unknown. Black swans are events that typically have a low probability of occurrence but ultimately have substantial consequences.[57] Theoretically, black swans, then are also known unknowns, but being known is problematic as the events are easily overlooked when conventional knowledge is guiding the search.

Summary

The probability of global catastrophes and human extinction is malleable. Scientists have estimated the probabilities of extinction-level events but the impacts of those events on humanity are largely within our ability to control. This chapter focused mainly on methods to help us anticipate events and actions that could be problematic for our future. Rigorous development of SCES can help us better understand where to focus our efforts to short-circuit those chains and also to shed light on whether we have exceeded the ethical threshold for the risk of extinction. Robust assessments of unintended consequences that could arise from emerging technologies and other human actions can help us move unanticipated unintended consequences into the anticipated unintended consequences bin and then allow us to take corrective actions. Table 4.2 presents my subjective assessments about how well humanity is implementing best practices to deal with risks and conducting SCES, UUCs, and UKUK assessments. As expected, overall, the assessment suggests humanity has to put more efforts into all of these areas with respect to extinction-level events. The assessment is barely improved with respect to global catastrophic risks. I would say, though, that entities in the business

Table 4.2 Assessment of How Well Humanity is Implementing Actions to Deal with Major Extinction and Global Catastrophic Risks

	With Respect to Extinction-Level Events	*With Respect to Global Catastrophic Risks*	*With Respect to Business Risks*
Best Practices	F	C	P
SCES Development	F	F	C
UUC Analyses	F	F	F
UKUK Analyses	F	F	F

Legend: P (pass) – things are fine; C (concerned) – we should be concerned; and F (fail) – we are completely failing.

sector actively engage in anticipation, planning, mitigation, and adaption in order to survive and thrive. However, it is not clear to me if they employ the additional three approaches. I hypothesize that the development of a global data repository will greatly help in these efforts.

Whole earth monitoring system

Effectively addressing the many problems facing humanity, including our own survival, requires a great deal of data. Thus, it is recommended that over time humanity build a whole earth monitoring system (WEMS). WEMS is one of two major systems proposed in this book. The other is called the Global Anticipatory and Decision Support System (GADSS), which is introduced in Chapter 7.

Figure 4.5 presents a schematic for such a system. Essentially, the system would monitor planetary systems and flows between systems that represent both the anthroposphere and the geosphere from the global scale down to the microbial scale, while also allowing for data to be collected from virtual environments found here on the earth or elsewhere. Building WEMS will require a tremendous amount of work. In the parlance of Perpetual Obligation # 11, which is presented in more depth in Chapter 6, this would be a

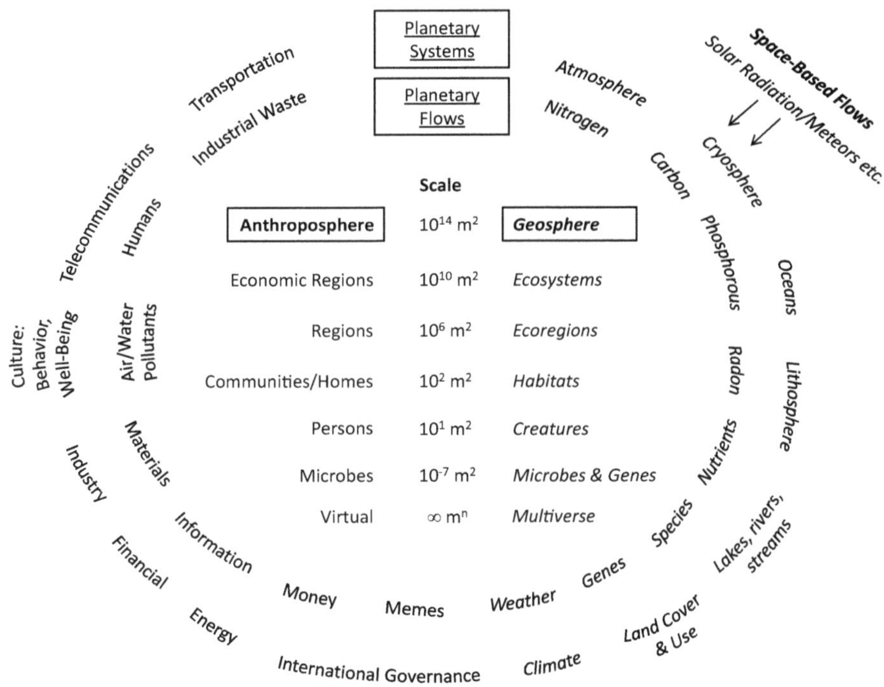

Figure 4.5 A Schematic Design for a Whole Earth Monitoring System

no-regrets, leave-no-unfinished-business project that would mark a major achievement in the history of humanity.

WEMS would essentially be the world's database. The data would be meticulously collected and quality assured. Appropriate metadata would be developed for each datum to facilitate reductionist and system of systems analyses. WEMS should not be confused with a being a world model or any other type of simulation model. On the other hand, models of every conceivable sort, from models as grand as global climate models to those that focus on modeling changes in regional land use, could draw inputs from WEMS. WEMS could also be expected to provide data for innumerable statistical analyses.

Some of these analyses could directly benefit the topics discussed in this chapter. For example, a great deal of data and analyses are needed to develop a rigorous understanding of the UUCs that could emanate from various events. Dori Stiefel and I have worked on a framework to help understand how pervasive UUCs might be with respect to emerging events. Pervasiveness can be measured by how many societal systems the event may influence. In other words, the more systems a technology or action can impact, the more UUCs will be created and the more severe will be the accumulated impacts of the UUCs. Our framework incorporates seven systems:[58]

(1) Social – culture, religion, and daily living;
(2) Technological/virtual – nanotechnology, biotechnology, information technology, cognitive technologies, energy and space technologies, and materials science;
(3) Economic – production, trade, and finance;
(4) Environmental – ranges from local ecosystems to various global systems (e.g., oceanic, atmospheric);
(5) Space – satellites, space stations, space transport, and other space-based systems;
(6) Political/military – encompasses domestic and international communities and relationships mediated through cooperation, negotiation, and military action;
(7) Human/health – physical and mental health of humans.

Figure 4.6 plots about two-dozen existing (in normal font) and emerging technologies (in bold font) by the number of systems they touch (*x*-axis) and our subjective estimate of the risk ensuing UUCs might pose to future generations (*y*-axis). One can discern a linear relationship between the number of systems impacted and the anticipated risk to future generations. It took us a fair amount of time to build our very limited data set. We are not able to track secondary or beyond systems impacts. We were not able to assess how the impacts of events coalesce, converge, and/or cascade. A WEMS would have allowed us to consider many more events and track UUCs in a much

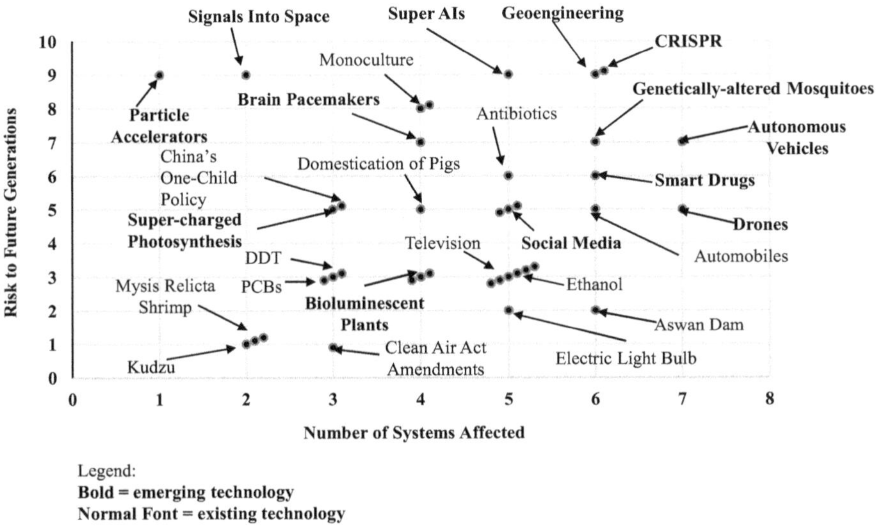

Figure 4.6 Risks to Future Generations From the Unanticipated Unintended Consequences of Selected Existing and Emerging Technologies

more sophisticated fashion. These results could then allow us to build a better predictive model that could be applied to emerging technologies and other events. The outputs from the predictive model would then guide our monitoring of WEMS for weak signals that any UUCs might be emerging.

WEMS could also be invaluable to resolving UKUKs. Creating new knowledge to resolve UKUKs could require an extraordinary amount of data mining. A new systemic phenomenon could be expected to be describable through correlations among previously unrelated variables. The ensuing system could emerge at any scale represented in Figure 4.5 and encompass any new combination of variables. Thus, the more data and variables that WEMS incorporates, the more comprehensive and exhaustive our search to resolve UKUKs could be.

WEMS is also needed to support other aspects of the futures-oriented blueprint outlined in this book. Very importantly, WEMS is needed to provide data needed to assess whether current generations are meeting perpetual obligations to future generations. I do apologize for referencing perpetual obligations as much as I do before they are actually presented in Chapter 6. As might be expected, one perpetual obligation is to prevent humanity from becoming extinct. WEMS' resources would be vitally important to anticipate the many extinction-level risks discussed in Chapter 3 and whether the 10^{-20} risk threshold is being met.

Another perpetual obligation is reducing the risks individuals face from involuntary environmental risks especially from air and water pollution and toxic chemicals in our food and indoor environments. Environmental health research, in general, is severely constrained in its funding compared to the task. For example, it is estimated that there are more than 80,000 toxic chemicals in use around the globe.[59] Only a relative handful of chemicals have been thoroughly researched for their impact on human health. Even in these cases, it is rare for there to be enough data from human sources or animal trials to firmly establish dose–response curves.[60] It is also rare that this research explores likely combinations of exposures to toxins.

Thinking long term, this research could be bolstered by efforts to create human exposomes. An exposome is a catalogue of every potentially toxic material that an individual is exposed to during their lifetime, including in the womb.[61] Figure 4.7 presents a very simple illustration of the concept. If exposomes were prospectively created for every human being and their health histories, and causes of death were captured, then data mining procedures could begin to establish the impacts on human health not only of exposures to specific chemicals but also to various combinations of chemicals. This is a huge task but accomplishable over the long term using WEMS.

Of course, WEMS cannot only be a data repository. It will need additional capabilities to make it useful to humanity. WEMS would have to have an exceptional geographic information system base. An intelligent front-end interface would be needed to assist users in linking data for specific analyses and then to most effectively display results. The interface would need to be multilingual and available to those with any impairment that would make standard interfaces difficult to use.

WEMS intelligent assistants (WEMSIAs) would be needed to help users interpret results. WEMSIAs would also help users spot the weakest of signals

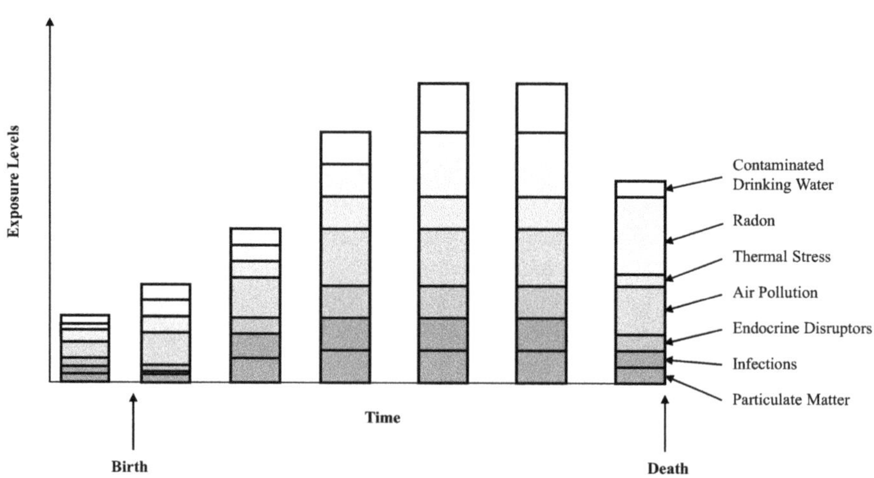

Figure 4.7 An Illustration of a Simple Exposome

that could reveal UKUKs. Additionally, WEMSIAs could help users develop SCES, first by identifying trends that form the backbones of the SCES and then to test for plausibility of conditions over time described within the scenarios. This functionality could also help users comprehend new systems dynamics and maybe even the emergence of new systems, which then could lead to new WEMS data collection requirements. Lastly, WEMS could serve as a platform for the convergence of systems of knowledge.[62]

Of course, a WEMS carries with it major risks for invasions of privacy and mis-use. I am aware that creating WEMS will require an unprecedented amount of cooperation and trust between nation states, which seems a very daunting challenge given the state of international relations in the first part of the twenty-first century. I do not have any easy solutions to propose. I do note that if humanity is up to the task of anticipating risks to its own extinction, it must also become mature enough to deal with the many trials and the unintended consequences of this project.

Broad and vastly open-minded programs are needed to identify and reduce the UKUKs that could be secretly troubling our futures. Addressing SCES, UUCs, and UKUKs requires a great deal of effort worldwide with the assistance of WEMS.

Singular chain of events scenarios: example

This chapter concludes with an example of what I imagine would be a typical SCES. This scenario is essentially an extended version of the scenario captured in Figure 4.3. I understand that this scenario is fanciful, maybe even bordering on science fiction. The probability that this scenario could actually occur is quite small. However, developing scenarios like this one discussed here can provide insights into the how's and why's of SCES. As discussed in depth in Chapter 6, if the world had a very large set of SCES developed, in total they could be used to develop more nuanced estimates of the probability of human extinction.

A singular chain of events[63]

Waves of change and resistance to change were sweeping the earth in the first decades of the twenty-first century. The Haves were riding the wave toward the Singularity, the promised land of perfect health, never-ending youth, and bountiful happiness.[64] Nanotechnology, biotechnology, information technology, and cognitive enhancements push the envelope before we die so we would not have to.

The Havenots, comprising about 90% of the rest of the human population and numbering in the billions, were suffering, as always. The Havenots lot included more than the destitute. This group also included the not so smart, the not so chic, and the religious zealots, who found solace in a cause, not in the competition of the capitalist game. The anger of this lot grew by the day, as the rules of the capitalist game kept changing in favor of the Haves[65] and as many of the Haves played on gullible Havenots for political support.[66]

For the most part, the Haves did not pay much attention to the Havenots. The Haves were optimistic that their new technologies would lead to utopia. Through their political muscle and consumer dollars, researchers and companies were hell-bent on inventing the technologies to make this new world a reality. They neglected basic environmental science and geological realities and took for granted that the food they ate and the water they drank would always be available. Energy efficiency, biodiversity, carbon sequestration, space exploration (along with precautionary inhabitation of other planets), and other important issues were deemed much less important than their own immortality. In fact, despite the calls by futurists, humanity had no presence in space; every single living human was confined to the earth. Anticipating very long lives, they robustly supported policies that promoted accumulation and protection of their wealth, which they intended to live on for many, many years.

The Havenots' focus on the Haves was decidedly pathological. The Havenots were obsessed with either becoming like the Haves or destroying the Haves. They could never quite develop the political leadership to effectively confront the Haves, as the most talented of the Havenots eventually became co-opted by the pseudo-meritocracy of the Haves. Neither the Haves nor the Havenots were paying attention to the big picture, the survival of all. Foresight was only considered in hindsight.

The first, explosive step along the path to human extinction was years in the making. In the mix were very familiar variables, oil, religion, economics, terrorism, and military might. Despite warnings from the small cadre of environmentalists, advocates, and futurists, the world had not been able to wean itself off oil, a key global economic resource that virtually had no ready substitute. The United States, a nation that rose to dominance on the back of oil, was particularly unable to lessen its dependence on oil. Oil prices were becoming too much of a drain on the economies of the importing nations and the money flowing to Arab countries was still funding terrorist activities around the world.

Efforts to move to a hydrogen economy had stalled. Russia and South Africa, the world's major producers of platinum for fuel cells, conspired to raise the price of raw platinum. National systems for the production, transport, and storage of hydrogen never materialized. Therefore, most of the few fuel-cell vehicles on the roads ran on natural gas and had onboard reformers to extract hydrogen from the natural gas. Production of gasoline from biomass, coal, and oil shale faced complaints from environmentalists on a number of issues. Neither consumers nor political leaders, slaves to public opinion polling of the ironically myopic Haves, were able to offer the leadership necessary to overcome the host of barriers to alternatives to oil.

Meanwhile, it was apparent even at the beginning of the twenty-first century that a military clash of major proportions was on the horizon when China's economy began to expand and soak up a growing percentage of the world's oil. China's initial energy strategy was to secure foreign oil resources. For many years, it tried to buy these resources, using its enormous trade surpluses gained from trade with the United States and the rest of the world. When this strategy proved futile, it decided that military action would be needed to secure future oil supplies.

Of course, this was not an innovative policy, as it was the policy that had been pursued by the United States, and implicitly (if secretly) backed by the Western European countries and Japan, for several decades. The United States intelligence community could easily track the growth of the Chinese military capabilities, including the construction of warships, aircraft carriers, and other forces required to conquer existing oil fields. In response, the United States prepared to launch a preemptive strike in the Middle East to secure its oil fields. The buildup of the Chinese military and potential U.S. responses were daily fodder for the incessant 24x7 news cycle.

The oil-producing countries in the Middle East only had one defense against the impending invasion; they threatened to destroy the oil fields with nuclear bombs to prevent the superpowers from again defiling their lands through another hostile invasion. Western intelligence agencies, led by the CIA, made their last monumental mistake when they assured their leaders that the threat was baseless. What they did not know was that Russia had "allowed" the theft of nuclear materials needed to build six nuclear bombs. Pakistan and North Korea provided the technology and weapons facilities. Had the Western nations spent the money to secure Russian nuclear materials years before, this could not have happened. In any case, Russia believed this was an excellent opportunity to reap a financial windfall, as it would then become the world's leading oil producer, while the other two countries wanted to bloody the noses of the United States and its Western European and Japanese allies.

On a cloudy morning in Washington DC, the President of the United States issued the order to launch the attack to take over the oil fields in the Middle East, once and for all. As the first soldiers parachuted down to the fields, the buried nuclear devices were set off in the major oil fields of Saudi Arabia and Kuwait. A significant fraction of the world's proven oil reserves was permanently destroyed in a matter of seconds. As expected, China launched a counterattack to try to secure the remaining oil fields in the Middle East. The subsequent warfare destroyed the remaining oil facilities and left a large percentage of the remaining oil fields in flames. In an act of monumental irrationality, these two nations also hunted down and destroyed oil tankers on the oceans, in ports, and even in canals in order to prevent their adversaries from securing even one more drop of oil.

The Russians and the rest of the oil-producing nations, from Venezuela to Nigeria, might have believed for a few hours that the value of their remaining oil resources would soar, along with their financial prospects. Unfortunately, a heretofore neglected terrorist group had been working under the radar to finish off the remaining oil reserves. In concert with the radicalization of religious movements, the environmental movement also continued to radicalize. Many in this movement had come to believe that the burning of fossil fuels was literally killing the planet earth. The most radical believed that humans should become extinct sooner than later, not to join their Lord sooner than later, but to benefit the remaining species on the earth.

These environmentalists also decried the frequent oil spills that despoiled beaches and killed wildlife and indigenous forms of life. Some well-meaning environmentalists worked on oil-eating bacteria to help clean up oil spills. In hindsight, it was readily predictable that before long, these bacteria would be weaponized by environmentalists. To try to forestall what they thought was going to be World War III (and maybe and hypocritically, to hasten the extinction of humans), they accelerated their development and deployment of their oil-field eating bacteria. A couple of days before the U.S. invasion of the Middle East, environmentalist operatives were able to disperse weaponized oil-eating bacteria into the flaming remains in the Middle East and all major oil fields around the world, including oil shale fields in Canada.

The world's first hint that a second even more destructive attack had been perpetrated on the oil fields came when the fires in the Middle East went out unexpectedly fast. The heat from the fires warmed the remaining oil resources below, which acted to accelerate the growth of the bacteria. This accelerated growth, combined with the exhaustion of oil in the ground from the fires, depleted the fuel for the fires in a matter of weeks. The fields were empty of oil. Inspectors discovered the bacteria by accident, when a few drops of goo from a field ate their petroleum-based gloves. Soon, reports were filed from elsewhere around the globe that oil-field eating bacteria were rapidly depleting the remaining oil supplies. Mutated forms of the bacteria were also consuming natural gas and many other petroleum-based products they encountered.

Once the realization that oil and natural gas would no longer be available to fuel the global economic machine, it took only a few months for most human institutions to unravel. Transportation systems came to a halt. Industrial production was slashed. People in Western countries could not get to work, and most had no jobs to go to. Stock prices tumbled. The Great Depression of the twenty-first century had begun.

The United Nations was particularly ineffective at forestalling continued war between the United States and China. Soon, countries around the world were joining the two sides. In anticipation of times of international unrest, the United States had developed the doctrine of mutually assured assassination (MAA) to keep foreign leaders in line. Any act of hostility would result in their immediate assassination. The threat was enforced by a triad of threats: stealth drone planes that continuously circled the whereabouts of foreign leaders, airborne stealth fighter planes that flew continuously off the coasts of major countries, and human agents placed in positions with access to the leaders. Of course, this U.S. policy initiated a race among other countries to implement their own MAA operations. In a very sad sign of the times, the lives of many world leaders were sometimes shadowed by 10–20 assassination protocols.

The multiple nuclear explosions, destruction of the oil tankers, and the impotence of the United Nations overwhelmed any rationality that the U.S. President had in applying her MAA doctrine, which was the last major decision of her life. As expected, the U.S. decision set off a series of assassinations. Within six months, over 150 national leaders had been assassinated, including the President of the United States. Within a year, most of the leadership of most countries had been assassinated, died at the hands of their own people, or had fled into exile.

Worldwide political and economic chaos ensued. There were mass migrations from the poor countryside into already overwhelmed and unhealthy cities. In Asia, many of these immigrants took their farm animals with them including chickens and ducks harboring novel strains of avian influenza. These birds had been genetically engineered by the West to resist a broad set of flu viruses to deal with just this eventuality. Unfortunately, the Western paradigm was still based on controlling nature, not coexisting with nature. As a result, what happened with bacteria – such as streptococcus – when antibiotics hastened the evolution of antibiotic-resistant bacteria happened again, this time with the evolution of a new super-strain of avian flu virus. It was no surprise that the infection spread rapidly through the Asian bird population, destroying an important food resource. It was also no surprise when the virus mutated again and spread to the human population.

Hospitals in many countries, dependent in so many ways on fossil fuels for plastics and energy, were not prepared to cope with the massive number of people sick with the virus. Previously developed flu vaccines had only a 20% protection rate. Several new vaccines developed within the next 1–2 years were also duds. Amplifying the destructiveness of the flu epidemic were additional diseases, including tuberculosis, measles, cholera, and malaria. These maladies flowed through populations already weakened by HIV/AIDS, diabetes, asthma, obesity, heart disease, and cancer (e.g., from smoking, air and water pollution, and destruction of the protective stratospheric ozone layer), who were largely unable to get maintenance or preventative care for their conditions.

No nation, except the United States and a few other developed countries, had functional, though depressed, public health systems. Within a year, just over two billion people died from infectious diseases, roughly one-third of the human population, about the same as the infamous Black Plague.

Unfortunately, unlike the aftermath of the Black Plague in the Middle Ages, human population continued to slide, or it should be said that the number of Havenots continued to decrease. Next up was climate change. For aforementioned reasons, the countries of the world had not stemmed the use of fossil fuels and, therefore, had not reduced the emissions of GHG into the atmosphere. Few technologies had been put in place to reduce GHG in the atmosphere or sequester the carbon elsewhere. It was as if the planet extracted revenge through withering droughts in Central China, Northern Africa, and North Central North America, deadly heat waves in Western and Central Europe, implacable sea-level rise in the Asia Pacific region, and apocalyptic storms worldwide. People were literally washed away down rivers and into oceans. Agricultural systems collapsed outside of the wealthy areas of the Haves, which were quickly becoming self-sufficient and hermetically sealed to the world of the Havenots. The built environment and urban infrastructures were pummeled. Another round of diseases, mostly mosquito-borne this time, ravaged the world's population. The developed world offered no safety net for the rest of the world. The largest losses of population were in Asia and Africa, closely followed by Central and South America. Within another thirty years, another billion people perished.

During the next handful of decades, the remaining humans failed to bond together to rebuild human civilization. In fact, just the opposite happened. Instead of conflicts between nations or even 'clashes of civilizations', deadly and widespread violence arose between the Haves and Havenots. At the outbreak of the unrest, the militaries of the world had been deployed to protect the wealth and property of the Haves. First as pandemics roiled the world and then as major economic systems collapsed, the viciousness and desperateness of the attacks of the Havenots against the Haves increased.

The military leaders had a choice: defend the Haves or become allies of the Havenots. The Haves had all the technological advantages (not only their life-prolonging technologies, useful if they could survive the chaos, but also their nanotechnologies, biotechnologies, limited but developing renewable energy technologies, and information technologies). The Havenots had strength in numbers. The majority of the military leaders whose forces were equipped with the most sophisticated weaponry and other advanced technologies made a devil's bargain with the Haves, security in exchange for the promised long-life and luxury.

A protracted period of violence ensued. Both the Haves and Havenots suffered substantial casualties. Eventually, the Haves and their superior military forces ended up in approximately 5,000 heavily defended enclaves (or lifeboats, from their point of view), with about 1,000 humans in each enclave. Most enclaves were former military bases, although many were former resort islands and other easily defensible haunts of the Haves. The enclaves brought to mind the walled cities of the Middle Ages. Unlike their feudal ancestors, they did not rely upon serfs living outside the walls for their food and materials. Because of their technological prowess, they were, after a period of transition, mostly self-sufficient. The poor and otherwise 'useless' and 'excessive' inhabitants of the enclaves, mostly lower ranking soldiers but also some weak Haves, were quickly evicted so as not to stress the resources of their systems, which would need to last for centuries.

During this period of violence, the Haves and the military systematically destroyed all advanced technologies outside of the enclaves. This was done so that the Havenots could not develop the capabilities to conquer the remaining enclaves. The military effectively destroyed the remaining energy-producing facilities (including the nuclear power plants), the electricity infrastructure, the worldwide telecommunications infrastructure, shipping and transportation facilities, and even dams and irrigation systems. The result of these attacks was that the Havenots on the outside had little or no technology, no concentrated energy resources, no information technology, no electricity, no water systems, and no advanced weapons. Agricultural productivity approached pre-industrial levels. Plants stressed by heat and drought failed to produce crops. Farm animals and plants regularly fell to agricultural diseases that had previously been preventable. Wild animal stocks were slaughtered with no thought about tomorrow. Accessible stocks of fish in the lakes and oceans were depleted. Leadership and new government structures never re-evolved; anarchy reigned. New pathogens circled the globe with astonishing speed. It was every man, woman, and child for themselves. Over the next century, the reduction in population was steady, and another two billion perished.

Catastrophic changes in the world's ecosystems coincided with the violence and also plagued the Havenots. In a mad scramble to keep themselves fed, the Havenots severely depleted the world's stocks of birds and mammals, big and small. This reduction in the number of insect predators led to an explosion in the numbers of destructive insects. Locusts and grasshoppers devastated remaining agricultural crops. In a particularly gruesome twist of fate, the depletion of mammals and birds also reduced the food supplies for mosquitoes around the world. As their predators were eliminated and as their food supplies dwindled, they began to viciously swarm individual humans who lacked shelter. Many did not survive the onslaught.

The Havenots and the Haves alike were killed by immense fires. The dramatic rise in CO_2 in the atmosphere and the expansion of ranges of temperate and tropical ecosystems promoted the accelerated growth of plant life all over the planet. Megatons of increased biomass respirated increasing levels of oxygen into the atmosphere. The bacteria that consumed the remaining oil and natural gas reserves also emitted substantial amounts of oxygen into the environment. Indeed, the level of oxygen in the atmosphere quickly began to approach 30%, from a level of about 21% at the turn of the twenty-first century. More plant materials, drought, and oxygen-rich air led to truly horrific conflagrations in North America, Europe, Northern Asia, Southern Africa, and Central and South America. Humans died directly in the fires and also died of asphyxiation if they were in the vicinity of the most massive fires.

Life in the enclaves became decidedly dystopian. The main problem was that no enclave possessed the critical mass of people, knowledge, and materials to maintain their technological base. Technology failed. In most cases, it was impossible to replace and/or fabricate new specialized chips and parts. Because the enclaves had destroyed the globe's telecommunications infrastructure to deny the Havenots the ability to easily organize, they were unable to communicate with other enclaves. The Haves continued to perceive the Havenots and the 'outside', disease-ridden world, to be a threat, although had they left their enclaves they would have known otherwise. This perception kept the Haves sequestered in their enclaves. Over the next 100 years, most of the enclaves collapsed from starvation or were eventually overrun by the Havenots, having failed like the Utopian communities of the eighteenth and nineteenth centuries.

A few enclaves, however, took a different path to extinction. This is because some Haves did achieve part of their vision of Utopia during the hell storm that surrounded them. They did achieve some measure of immortality. In a handful of enclaves, there were Haves who were actually a couple of hundred years old. But they had not planned on the destruction of the rest of the world and their technologies were riddled with bugs.

It was imperative that these Haves strictly control their population. Despite their weaponry, they were essentially trapped in their enclaves. The outside world was disease ridden, chaotic, dangerous, and empty of valuable resources. They had no survival skills beyond their advanced technologies. They could not survive outside of the enclaves. Controlling their population meant that the births needed to be well planned and limited in number, especially since their numbers had been swollen with the ranks of the military.

The major flaw in this strategy is that these Haves, who desperately wanted to be immortal, basically achieved this goal. Through enhanced nutrients, key replacement organs, and medical nanotechnologies, they were able to keep their bodies in excellent condition. They were not afflicted with heart disease or cancer or obesity or diabetes. Their lives within the enclaves were rather safe because Havenots found the risks not worth the effort of confronting these small but deadly enclaves. The Haves did not travel at all nor have many on-site accidents. They were not murdered in the streets, although inevitably some were killed during disputes in their enclaves. They did not commit suicide; they were constitutionally incapable of taking their own lives, having committed themselves to immortality. After a while, the turnover in the enclaves fell to close to zero. No one died. And the enclaves could not afford to allow new births. These Haves were not too worried. After all, they had time on their side, right?

However, as time went on, these super-elders lost the ability to reproduce naturally. The eggs in the female's ovaries aged and could not be rejuvenated. Also, frozen eggs and sperm turned out to have much shorter shelf lives than had been thought. To reproduce, that left cloning. Although advances in cloning had been impressive, problems with human cloning had not been overcome simply because the practice had been banned by most countries at the beginning of the twenty-first century and had been taboo in the enclaves for most of the time. However, these Haves decided to try to clone humans though they lacked skilled scientists to oversee this process. The results were disastrous. Miscarriages were the most common results. Many fetuses that came to term died shortly after birth. Most were aborted, those that were allowed to go to term died minutes after being born. The very few that lived further were afflicted with cognitive deficiencies, deformed limbs, and, tragically, were infertile. The attempts at cloning were rapidly abandoned.

Another problem that these Haves did not anticipate was the psychological aspects of aging. The minds of these very old people were slowly becoming completely dysfunctional. Of course, they did not suffer from Alzheimer's or Huntington's or Parkinson's diseases. They had genetic tests for these maladies and could prevent or treat these diseases without much effort or risk. What they did suffer from was system overuse and overload. Too many memories over too many years were leading to inefficiencies in memory retention and organization. Sleep no longer was sufficient to help keep their minds organized. As their collective capabilities were eroding at about the same rate, these Haves were unable to recognize what was happening to them. Because of this creeping functional senility, they were also increasingly unable to maintain their other technologies in tip-top shape. Plans to move out of the enclaves vanished. Pictures of health, they were going mad down the path to extinction. Eventually, even these more resilient enclaves perished as their diminished mental capabilities proved insufficient to keep themselves alive.

When the last enclave fell, there were around 500 million Havenots left on earth. Then, what was once referred to as northwest Wyoming exploded in the largest volcanic eruption the earth had witnessed in the past 20 million years. The eruption was 10,000 times the size of the St. Helens eruption. The soot pushed up into the atmosphere severely blocked out the sun everywhere on the earth for several years. Plant life suffered due to the reduction in photosynthesis. Much like what happened several million years ago to the dinosaurs, the number of humans on the earth dropped down to the mere thousands.[67]

The remaining hunter-gatherer Havenots were exceedingly resourceful. Many were able to scrape by, living in caves, or building shelter from rubble and scavenging for food and water. They had been able to deal with the hell of climate change and seemed poised to deal with this new round of precipitous cooling. Unfortunately, a final sequence of events on a geological scale would soon seal their fate.

The Havenots were a very unlucky lot. Weakened from disease, malnutrition, and inbreeding, they were also becoming very lethargic, light-headed, and disoriented. The shortness of breath was the key symptom explaining this new malady. You see, they were beginning to suffocate because the oxygen levels in the atmosphere had dropped below 20%.

What had happened to the oxygen? The conflagrations had drawn a great deal of oxygen out of the atmosphere. The remnants of civilization were also oxidizing. Old bridges, steel buildings, and especially billions of metal automobiles, trucks, motorcycles, and signs were rusting and rapidly sucking oxygen out of the atmosphere. As the Havenots did not have the technologies in place to produce their own oxygen, they suffered from oxygen deprivation *en masse.*

Cooling continued to worsen. All the negative feedback effects were in place: severe reduction of sunlight, loss of plant life, reductions in greenhouse gases in the atmosphere, increased radiative cooling, further loss of plant life, further reductions in greenhouse gases, etc. The Arctic Ocean, already refrozen, started its march southward. The Antarctic ice fields rapidly expanded. Glaciers, which had reappeared on the mountain-tops, quickly moved toward the valleys. As a consequence, sea levels dropped precipitously worldwide.

As a result, enormous amounts of rock now stood bare to the elements. Land scrapped clear due to erosion from floods and storms was not revegetated and was also exposed to the elements. These rocks, along with the husk of human civilization, oxidized, drawing ever more oxygen out of the environment. Indeed, large areas of the earth began to resemble the Red Planet, Mars. All remaining aerobic species not only faced a life deprived of sufficient oxygen, but they now faced the prospect of asphyxiation. Within another couple of hundred years, the oxygen in the atmosphere dropped below 19.5%, on its way to a low of 15%. The last human took her last breath in Southern Africa, just like the last Gorgon did over 250 million years ago.[68]

Notes

1 B. E. Tonn and D. MacGregor. 2009. A Singular Chain of Events. *Futures,* 41, 10, 706–714.
2 T. Maher and S. Baum. 2013. Adaptation to and Recovery from Global Catastrophe. *Sustainability,* 5, 4, 1461–1479.
3 See Peter Schwartz's 1996 book, *Art of the Long View* (Doubleday, New York), for a wonderful introduction to scenario writing.
4 Intergovernmental Panel on Climate Change. 2000. *Emissions Scenarios – Report.* Nebojsa Nakicenovic and Rob Swart (Eds.). Cambridge University Press, Cambridge, 570. Available from Cambridge University Press, The Edinburgh Building Shaftesbury Road, Cambridge CB2 2RU ENGLAND.
5 International Energy Agency. 2019. *World Energy Outlook 2019.* IEA, Paris. Retrieved from www.iea.org/reports/world-energy-outlook-2019
6 www.millennium-project.org/projects/workshops-on-future-of-worktechnology-2050-scenarios/
7 CEs have been researched and developed for much more specific and limited contexts. For example, the field of probabilistic risk assessment (PRA) was

developed to study, identify, and develop responses to short-circuit cascading failures through the systems of nuclear power plants. Probabilistic Risk Assessment. 2019. *Wikipedia*, Wikimedia Foundation, March 14. Retrieved from en.wikipedia.org/wiki/Probabilistic_risk_assessment.

8 M. W. Shelly. 1826. *The Last Man.* Edited with an Introduction by Anne McWhir. Broadview Literary Texts, Ontario, Canada.

9 B. E. Tonn and J. Tonn. 2009. A Literary Human Extinction Scenario. *Futures*, 41, 760–765.

10 M. D. Paley. 1993. The Last Man: Apocalypse Without Millennium. In Audrey A. Fisch, Anne K. Mellor, and Esther H. Schor (Eds.). *The Other Mary Shelley: Beyond Frankenstein.* Oxford University Press, New York. 107–123.

11 D. Morgan. 2009. World On Fire: Two Scenarios of the Destruction of Human Civilization and Possible Extinction of the Human Race. *Futures*, 41, 10, 683–693.

12 T. Lopes, T. Chermack, D. Demers, K. Medhavi, B. Kasshanna and T. Payne. 2009. Human Extinction Scenario Frameworks. *Futures*, 41, 10, 731–737.

13 P. Carpenter and P. Bishop. 2009. The Seventh Mass Extinction: Human-Caused Events Contribute to a Fatal Consequence. *Futures*, 41, 10, 715–722.

14 W. Bainbridge. 2009. Demographic Collapse. *Futures*, 41, 10, 738–745.

15 C. Jones. 2009. Gaia Bites Back: Accelerated Warming. *Futures*, 41, 10, 723–730.

16 F. Goux-Baudiment. 2009. Tomorrow Will Die. *Futures*, 41, 10, 746–753.

17 B. E. Tonn and D. MacGregor. 2009. A Singular Chain of Events. *Futures*, 41, 10, 706–714.

18 A robust version of this scenario is presented at the end of this chapter.

19 This section is adapted from B. E. Tonn and D. Stiefel. 2018. Unintended Consequences of Scientific and Technological Purposive Actions. *World Futures Review*, First Published August 1.

20 R. K. Merton. 1936. The Unanticipated Consequences of Purposive Social Action. *American Sociological Review*, 1, 894–904.

21 B. W. Arthur. 2009. *The Nature of Technology: What It Is and How It Evolves.* Simon and Schuster, New York.

22 R. K. Merton. 1936, ibid.

23 R. Vernon. 1979. Unintended Consequences. *Political Theory*, 7, 1, 57–73.

24 M. Cherkaoi. 2007. *Good Intentions: Max Weber and the Paradox of Unintended Consequences.* Bardwell Press, Oxford.

25 K. Popper. 2002. *The Poverty of Historicism.* Psychology Press, New York.

26 P. Baert. 1991. Unintended Consequences: A Typology and Examples. *International Sociology*, 6, 2, 201–210.

27 C. Mazri. 2017. (Re) Defining Emerging Risks. *Risk Analysis*, 2053–2065.

28 W. McDermott. 1993. Of Unexpected and Unintended Futures. *Futures*, 25, 9, 997–1006.

29 R. Jackson and J. Salzman. 2010. Pursuing Geoengineering for Atmospheric Restoration. *Issues in Science and Technology*, 26, 67–76; H. Lu, Y. Zhang, Y. Guo, D. Zhu, and A. Porter. 2014. Four-Dimensional Science and Technology Planning: A New Approach Based on Bibliometrics and Technology Roadmapping. *Technological Forecasting and Social Change*, 81, 39–48; C. Song, D. Elvers, and J. Leker. 2017. Anticipation of Converging Technology Areas–A Refined Approach for the Identification of Attractive Fields of Innovation. *Technological Forecasting and Social Change*, 116, 98–115; H. A. Linstone. 2001. TF/TA: New Driving Forces. *Technological Forecasting and Social Change*, 68, 3, 309–313.

30 B. E. Tonn and D. Stiefel. 2013. Evaluating Methods for Evaluating Existential Risks. *Risk Analysis*, 33, 10, 1772–1787.

31 J. Shapiro. 2001. *Mao's War against Nature.* Cambridge University Press, New York.

32 K. Kelly. 2011. *What Technology Wants.* Penguin Books, New York.

33 A. Webb. 2016. *The Signals Are Talking: Why Today's Fringe Is Tomorrow's Mainstream.* Public Affairs, New York.
34 D. Roos, J. Womack, and D. Jones. 1991. *The Machine That Changed the World: The Story of Lean Production.* Harper Perennial, New York.
35 D. Murphy and C. Hall. 2011. Energy Return on Investment, Peak Oil, and the End of Economic Growth in Ecological Economics Reviews, Robert Costanza, Karin Limburg & Ida Kubiszewski, Eds. *Annals of the New York Academy of Science,* 1219, 52–72.
36 J. Rubin. 2016. Connected Autonomous Vehicles: Travel Behavior and Energy Use. In Gereon Meyer and Sven Beiker (Eds.), *Road Vehicle Automation 3.* Springer International Publishing, New York, 151–162.
37 Appendix B provides an extended extinction scenario titled *Death By Autonomous Vehicle.*
38 A. Wise, K. O'Brien, and T. Woodruff. 2011. Are Oral Contraceptives a Significant Contribution to the Extrogenicity of Drinking Water? *Environmental Science and Technology,* 45, 1, 51–60.
39 U.S. [United States] Census Bureau. 2015. U.S. [United States] Census Bureau History: Public Broadcasting. Report, Washington, DC: U.S. [United States] Census Bureau.
40 N. Minow and C. LaMay. 1996. *Abandoned in the Wasteland: Children, Television, and the First Amendment.* Hill and Wang, New York, 3.
41 R. Putnam. 2000. *Bowling Alone: The Collapse and Revival of American Community,* vol. 6. Simon & Schuster, New York.
42 E. Kabir, M. Rahman, and I. Rahman. 2015. A Review on Endocrine Disruptors and Their Possible Impacts on Human Health. *Environmental Toxicology and Pharmacology,* 40, 1, 241–258.
43 United States Environmental Protection Agency. 2017. Overview for Renewable Fuel Standard, June 7. Retrieved January 9, 2018, from www.epa.gov/renewable-fuel-standard-program/overview-renewable-fuel-standard.
44 J. Griffin. 2013. U.S. [United States] Ethanol Policy: Time to Reconsider? *The Energy Journal,* 34, 4, 1–25.
45 S. Holland, J. Hughes, C. Knittel, and N. Parker. 2013. *Unintended Consequences of Transportation Carbon Policies: Land-Use, Emissions, and Innovation.* Working Paper 19636, National Bureau of Economic Research, Cambridge, MA.
46 The Economist. 2006. Ethanol: The Law of Unintended Consequences. October 25.
47 J. Palca. 2017. EPA Approval of Bacteria to Fight Mosquitoes Caps Long Quest. *National Public Radio,* November 8.
48 J. Diamond. 1997. *Guns, Germs, and Steel: The Fates of Human Societies.* W.W. Norton, New York.
49 D. Rumsfeld. 2002. Department of Defense News Briefing, February 12. Retrieved August 8, 2012, from www.defense.gov/Transcripts/Transcript.aspx?TranscriptID=2636
50 The Institution for Science Advancement. 2016. The Unknown Unknowns: Plato's Allegory of the Cave, March 18. Retrieved from ifsa.my/articles/the-unknown-unknowns-platos-allegory-of-the-cave.
51 N. Myers. 1995. Environmental Unknowns. *Science,* 269, 358–360.
52 J. Ito and J. Howe. 2017. Emergent Systems Are Changing the Way We Think. *The Aspen Institute,* February 2. Retrieved from www.aspeninstitute.org/blog-posts/emergent-systems-changing-way-think/
53 B. Fischhoff and D. MacGregor. 1982. Subjective Confidence in Forecasts. *Journal of Forecasting,* 1, 155–172.
54 Riel Miller. 2010. Futures Literacy – Embracing Complexity and Using the Future. *Ethos Civil Service College,* 10, 23–28.

55 M. Harrysson, et al. 2014. *The Strength of 'Weak Signals'.* McKinsey & Company, McKinsey Quarterly, February. Retrieved from www.mckinsey.com/industries/technology-media-and-telecommunications/our-insights/the-strength-of-weak-signals#

56 S. Shostak. 2018. NASA Scientist Says Space Alien Search Should Be More 'Aggressive'. *SETI Institute,* December 12. Retrieved from www.seti.org/nasa-scientist-says-space-alien-search-should-be-more-aggressive.

57 N. Taleb. 2007. *The Black Swan: The Impact of the Highly Improbable.* Random House, New York.

58 B. E. Tonn and D. Stiefel. 2018. Unintended Consequences of Scientific and Technological Purposive Actions. *World Futures Review,* First Published August 1.

59 www.nrdc.org/issues/toxic-chemicals

60 https://en.wikipedia.org/wiki/Dose – response_relationship

61 www.cdc.gov/niosh/topics/exposome/default.html

62 B. E. Tonn, M. Diallo, N. Savage, N. Scott, P. Alvarez, A. MacDonald, D. Feldman, C. Liarakos, and M. Hochella. 2013. Convergence Platforms: Earth-Scale Systems. In M. Roco, W. Bainbridge, B. E. Tonn, and G. Whitesides (Eds.), *Convergence of Knowledge, Technology and Society: Beyond Convergence of Nano-Bio-Info-Cognitive Technologies.* Springer International Publishing, Dordrecht, 95–137.

63 This scenario was first published in B. E. Tonn and D. MacGregor. 2009. A Singular Chain of Events. *Futures,* 41, 10, 706–714.

64 R. Kurzweil and T. Grossman. 2004. *Fantastic Voyage: Live Long Enough to Live Forever.* Rodale, Emmaus, Pennsylvania; R. Kurzweil. 2005. *The Singularity Is Near: When Humans Transcend Biology.* Penguin Books, New York.

65 Z. Sardar and M. Davies. 2002. *Why Do People Hate America?* Icon, Cambridge.

66 T. Frank. 2004. *What's the Matter With Kansas? How Conservatives Won the Heart of America.* Metropolitan Books, New York.

67 S. Ambrose. 1998. Late Pleistocene Human Population Bottlenecks, Volcanic Winter, and Differentiation of Modern Humans. *Journal of Human Evolution,* 34, 6, 623–651.

68 P. Ward. 2004. *Gorgon: Paleontology, Obsession, and the Greatest Catastrophe in Earth's History.* Penguin Books, New York.

5 Why we should care about future generations

Introduction

Science has shown us that life on the earth must be understood within very long-time frames. The universe is about 13 billion years old. The earth formed about 4.5 billion years ago. Life evolved on the earth 3.8 billion years ago. Hundreds of millions of years of plate tectonics have changed the landscape of the earth. Fossils from the Cambrian Explosion 500 million years ago deposited in the sea can now be found in a mountain range in Canada.[1] Dinosaurs went extinct 65 million years ago. Homo sapiens appeared on the scene over the past couple hundred thousands of years. Such information was not available to our distant ancestors.

For millennia, the lives of our hunter-gatherer forbearers did not change much if at all. The next year would be much like the past year. The future was experienced through seasons, and movement of food sources from one place to the next. Mention of concerns about future generations is sparse over the course of human history, though there are notable exceptions. One example is the seventh Generation Principle that dates to the Great Law of Haudenosaunee, the founding document of the Iroquois Confederacy. This principle holds that important decisions made by the Iroquois Nation should be made with concern for the next seven generations.[2]

Robert Heilbroner explains that futures as an active psychological construct began to emerge with the Industrial Revolution.[3] New technologies were rapidly changing not only the quality of life but also lifestyles, work, and economies, the built environment, and even warfare. Progress became a theme to rally around, or fear. In either case, thinking about the future became important because the future was very unlikely to resemble the past.

It also needs to be stated that modern generations have a greater capacity to anticipate futures. We can predict the tides with high accuracy and the weather with increasing accuracy. Lending institutions provide mortgages that extend for decades. Some property leases reach a century in length. Our science allows us to consider extraordinarily long-time frames and imagine how future worlds could be different from ours. Chapter 2 illustrated how we can use our imaginations to journey through the future. The future promises to be extraordinarily interesting and challenging, as life always seems to be.

Chapter 3 illuminated the risks that may render our species extinct. Chapter 4 argued that our existence is, to a very large measure, our own responsibility. Existing extinction-level risks, such as nuclear war and climate change, appear to present significant and hopefully not insurmountable hurdles to demonstrating our ability to act rationally to protect ourselves and future generations.

Concerns about these risks surfaced in the futures studies literature under the rubric of *obligations to future generations*. As noted in Chapter 1, this literature emerged in the 1970s and 1980s in response to threats from nuclear war as well as the very long-term burden that nuclear wastes from nuclear power plants placed upon future generations. Futurists, philosophers, and others wrote extensively on the topic of future generations during this time frame.[4]

Futurists and others have worked to set out both reasons why we should care about future generations and what our obligations should be to future generations. For example, one reason we should care about future generations is that caring for future generations is a decisive intuition.[5] Much of the literature on obligations to future generations is framed in terms of sacrifice. Why should humanity care about the quality of life of its future generations? Why should current generations 'sacrifice' pursuit of short-term, intra-generational goals of wealth and political power, or simply having a good time, for future generations? A goal of this chapter is to argue that the sacrifice frame is actually not the correct frame for this discussion. In fact, concern for future generations should be seen as a duty and an intrinsic part of having the opportunity to be alive in the first place.

Two eminent futurists, Wendell Bell and Richard Slaughter, have each developed thoughtful statements on these subjects. Bell[6] puts forth these seven declarations:

- A concern for present people implies a concern for future people;
- Thought experiments in which choosers do not know to which generation they belong rationally imply a concern for both present and future people;
- Regarding the natural resources of the earth, present generations have no right to use to the point of depletion or to poison what they did not create;
- Past generations left many of the public goods that they created not only to the present generation but to future generations as well;
- Humble ignorance ought to lead present generations to act with prudence toward the well-being of future generations;

- There is a *prima facie* obligation of present generations to ensure that important business is not left unfinished; and
- The present generation's caring and sacrificing for future generations benefits not only future generations but also itself.

In a similar vein, Slaughter[7] makes these points with respect to caring about future generations:

- The human project is unfinished;
- Caring for future generations is ethically defensible;
- We are partly responsible for the dangers to their well-being;
- The global commons has been compromised by human activity and restorative actions are necessary;
- To not care about future generations diminishes us; and
- Caring for future generations is a cultural force that is valuable now and for the foreseeable future.

One can see that 'why current generations should sacrifice for future generations' is explicitly addressed by both authors. Bell's seventh point states that sacrifice benefits both current and future generations. Slaughter's last two points essentially make the same case.

One can see that both lists also contain reasons why we should care about future generations and examples of obligations to future generations. For example, Bell's statement "A concern for present people implies a concern for future people" addresses the 'why' question whereas "Regarding the natural resources of the earth, present generations have no right to use to the point of depletion or to poison what they did not create" essentially represents an obligation or course of action. Similarly, Slaughter's statement "To not care about future generations diminishes us" pertains to 'why' and "The global commons has been compromised by human activity and restorative actions are necessary" is definitely stated as an obligation.

One goal of the next two chapters is to disentangle the 'why' and the obligational statements. Additional goals are to expand our notions of what it is we need to care about and our reasons for caring. I argue that it is necessary but not sufficient to only care about future generations. There needs to be something more fundamental than choosing or not to sacrifice, to making this an individual-by-individual decision. There needs to be something more metaphysical about caring for future generations.

What this something is, I believe, is the additional need to care about *the journey*. Thus, the central philosophical question is not why we should care about future generations. The central question is why we should care about protecting, preserving, maintaining, and fostering humanity's journey through time and space. Not only does the idea of a journey resonate well with human myth and psychology, but it also fits better with our discussions of human extinction. Referring to Figure 4.4 again, the cone of possible paths through time shows paths that are cut short, meaning that sometime along those paths the journey ends. All of humanity's achievements, wealth, hopes, and dreams die at that point. Therefore, it is also necessary to explicitly embrace the concept of the journey and also care about the quality of the journey, which then also brings in the concepts of equity and sustainability.

On my desk at home, I keep one of my dearest possessions, a small block of wood from a Cyprus tree. It has The Imperial Chrysanthemum Seal of Japan on the back. This block of wood came from a Shinto Shrine called Ise Jingu that was dismantled in 1993. According to tradition, the Shrine is dismantled and rebuilt every 20 years using traditional techniques. This Shrine's history dates back to the year 690 and has been rebuilt over 60 times. A nearby forest provides the wood to rebuild the Shrine. This is an elegant futures-oriented tradition, because by custom the wood used to rebuild the Shrine needs to be several hundred years old. This means that the forest managers must plan several centuries ahead in order to have a ready supply of wood. This tradition symbolizes to me the essence of the journey and its relationship to futures.

My treatment of 'why we should care about future generations' begins by examining Bell's and Slaughter's most relevant statements from a values context. Six questions posed by Socrates are used to shape this first discussion. The second section addresses the concept of the journey. The journey is one of humanity's most important archetypes, but there are many types of journeys. It is important to understand what type of journey humanity is on. This part also addresses how both protecting the journey and caring for future generations fit naturally with fundamental human psychology. The third part takes a rationalist view toward this question: why should we care about protecting the journey of future generations through time and space? This discussion will be anchored by the concepts of the veil of ignorance and original position promoted by John Rawls as useful in thinking through aspects of a just society. This section also addresses the need for the concepts of equity and sustainability in our mission statement. The fourth section argues that humanity has an obligation to all other life on the earth to maintain the journey and find new homes beyond our solar system.

Caring for future generations should be a deeply held value

In his book *Six Questions of Socrates*, Christopher Phillips asks participants in Socrates Cafés held around the world to discuss their values in the context of these six questions: What is Virtue? What is Moderation? What is Justice? What is Good? What is Courage? What is Piety?[8] Phillips then addresses whether the excellence needed to live life in accordance with these values is possible in modern times. Let's use these questions as a starting point for exploring why caring about future generations should be a deeply held value. I am not going to argue that these six values are all encompassing.

However, they are sufficiently representative for our purposes. I offer these revisions of each question to fit our context.

- *Is caring about future generations virtuous?* Café participants frequently offered that virtue means having a higher purpose in life than just focusing on one's self. In this light, caring about future generations is virtuous. Virtue is explicit in Bell's statement that "a concern for present people implies a concern for future people."
- *Is moderation a value that benefits future generations?* This value was interpreted to mean finding a middle way through life. To be modest. To avoid extremes. Moderation in consumption of non-renewable resources and emission of greenhouse gases would most certainly benefit future generations. Moderation is needed to support sustainability goals. I think that this statement by Bell is consistent with this value: "humble ignorance ought to lead present generations to act with prudence toward the well-being of future generations."
- *Is caring for future generations just?* It is easy to confine the concept of justice to the righting of wrongs perpetrated by people with power on those who lack power. In a deep futures context, the people with power are current generations. Those without power are future generations. Bequeathing our future generations with horrific conditions or even preventing them from ever existing seems to embody injustice. As Slaughter notes: "We are partly responsible for the dangers to their well-being."
- *Is caring for future generations good?* Being good can be interpreted as doing things for others before doing things for one's self. One participant noted that being good is when you act in the interest of others even when no one is looking. Caring for future generations is quintessentially good because future generations cannot personally convey their gratitude for the goodness of current generations. I think that this statement from Bell fits well with this value: "The present generation's caring and sacrificing for future generations benefits not only future generations but also itself."
- *Does it take courage to care for future generations?* Courage is not reserved only for 'heroes'. Courage is doing the right thing when it is difficult to do so, when doing so may result in harm to yourself, physically, mentally, and emotionally. Caring for future generations in a modern society that glorifies individualism, consumerism, and short-term win–lose thinking seems to require courage. It also takes courage to directly face the question of human extinction and the complexities of unintended consequences and unknown unknowns. I do not believe that any of Bell's and Slaughter's statements directly address courage, but maybe this statement of Slaughter's could bolster people's inclinations to care for future generations: "Caring for future generations is a cultural force that is valuable now and for the foreseeable future."

- *Is it pious to care for future generations?* Piety was interpreted by many Café participants as having reverence and respect for fathers, elders, ancestors, and religion. I believe that it is also pious to have reverence and respect for future generations. Past generations sacrificed to benefit current generations. Current generations should do the same. As Slaughter notes: "To not care about future generations diminishes us."

My own view, then, is that caring for future generations is consistent with these six values addressed by Socrates. Individuals can achieve excellence by living lives that exemplify these six values. It is also clear that these six values are consistent with the reasons why we should care about future generations put forth by Bell and Slaughter. Interestingly, responses provided by the participants in the Socrates Cafés did not reference futures or obligations to future generations. However, Phillips offers this assessment on excellence at the end of their book:

> I think an excellent individual and an excellent civilization do share certain attributes: They are forward-looking. They are cognizant of how their actions impact others, not just today, but in coming generations, and strive to act in ways that will enhance the lives of individuals and societies not just of today, but also of the futures – and not just in the next one or two or five generations, but the next hundred and thousand and ten thousand generations.[9]

Journey of self-actualization

In the broadest of perspectives, the journey is not just a metaphor for life, it is an intrinsic characteristic of life on earth. Taken in all its forms on the earth, living organisms have an extraordinarily tenacious will to survive, to "keep on truckin" the words of the Grateful Dead.[10] How else can one explain the ability of life to bounce back from five previous mass extinctions, the thermophilic bacteria that live in the extreme heat of hot springs, the life found seven miles deep in the oceans?[11] How does one explain animals like the tardigrades, tiny creatures so hardy that they were recently sent to the moon and could be expected to survive if they are ever rehydrated?[12] Finally, how does one explain the inherent drive of evolution to diversify, to enter into symbiotic and synergistic relationships in every ecosystem on the planet, without recourse to the ontological concept of the journey?

Of course, humans are tightly interwoven into life's journey through time. The concept is deeply wired into our collective unconsciousness, according to Carl Jung.[13] More specifically, Jung identified the hero's journey as one of our most basic psychological archetypes.[14] Jung wrote "*The hero's main feat is to overcome the monster of darkness: it is the long-hoped-for and expected triumph of consciousness over the unconscious.*"[15]

Not surprisingly, the journey is a central theme in human myths, where the concept of myth in this context can be interpreted as traditional stories designed to help people to better understand their reality. Karen Armstrong explains that myth is often used to confront the experience of death and fear of extinction.[16] Joseph Campbell found that The Hero's Journey is a myth that transcends culture and time.[17] Joseph Campbell describes the hero's journey in his 1949 work *The Hero with a Thousand Faces* as: *A hero ventures forth from the world of common day into a region of supernatural wonder: fabulous forces are there encountered and a decisive victory is won: the hero comes back from this mysterious adventure with the power to bestow boons on his fellow man.*[18] The myths that weave through our cultures are not run-of-the-mill journeys. They are epic journeys, maybe worthy as plots for superhero movies. Our real futures journey is also heroic – surmounting climate change, super-volcanoes, gamma ray bursts, all the anticipated and unanticipated risks, and all the unknown unknowns.

The hero's journey is but one type of journey. Other types of journeys include the epic journey, the quest, rags to riches, a find yourself wandering, the tragedy, rebirth, vacation, to hell and back, and romantic. What kind of journey is humanity on? What kind of journey ought all generations be concerned about protecting?

It is tempting to characterize humanity's journey through space and time as a hero's journey. After all, it does require a heroic attitude to take on challenge after challenge not only to prevent our own extinction but also to prosper during the journey. And, at least at this point in time, humanity is a reluctant hero, not willing to immediately take on the responsibility for the heroic challenges of the journey.

However, upon reflection, humanity is not on a hero's journey. The hero is too much an individual and the experience of the hero is too individualistic. Also, humanity's journey ought not to be framed as fight against evil, unless framed as a fight between our own bright and dark sides. This is because the universe is not intrinsically evil. Finally, our survival does not depend upon traveling into hell, achieving enlightenment and victory, and returning as a heroic figure. Humanity's reluctance to be the hero is not borne out of modesty and humility, but out of ignorance and an unwillingness to be responsible, neither quality intrinsic to the hero. Currently, humanity is playing the narcissist (i.e., grandstander bully) and the coward.

Additionally, I would not say that humanity is on a quest, as we have no specific goals set out for our journey at this point in time. Our goals are too infinitesimally myopic to represent a noble quest that could require millennia to achieve. There have been and certainly will be tragic aspects to our journey, but characterizing our journey as a tragedy seems way too dark and pessimistic. On the flip side, it seems less than mature to romanticize our journey through time and space.

Humanity is undoubtedly on an epic journey, taken to extreme over millions of years with the expectation that we will colonize the galaxy, in part to avoid extinction from the death of our sun. We are also, in a sense, still wandering through time and space, or maybe better said, we are still actively exploring. Certainly, achievement and exploration, being driven by curiosity, are all components of our journey. On the other hand, we cannot expect to experience epic achievements on humanity's scale within any generation or individual's lifetime. For example, achieving cost-effective production of electricity by fusion will be an epic human achievement, but this particular quest has already spanned several generations of humans and it could well be that many more generations will need to contribute till that achievement is reached. The same can be said for fully colonizing Mars or even a planet outside of our solar system or to sending and successfully hearing back from a probe about conditions on a potentially inhabitable planet. Thus, the journey is more measured, shall we say an act of moderation that requires commitment without promise of grand achievement or fame at any point in time for any specific individual.

The concepts of achievement, curiosity, patience, and commitment are characteristics of mature and self-fulfilled individuals. So, maybe it is best to describe humanity's journey as one of *self-actualization*, to borrow a term from psychology. The concept was first proposed by Kurt Goldstein and then popularized by Abraham Maslow.[19] Carl Rogers also is known for using the term.[20] Applied to an individual, self-actualization can be understood as the full realization of one's potential, where potential has several dimensions, including social, intellectual, and creative. A self-actualized individual is also curious and able to indulge in their curiosities. Self-actualization is at the top of Maslow's hierarchy of needs.[21] Individual lives, then, can be seen as journeys toward self-actualization.

Humanity's journey through time and space can also be seen as a *journey of self-actualization*. To survive into the distant future will not only be an astounding intellectual achievement but will also require phenomenal creativity. The journey will also indulge humanity's curiosity about its own future and about aspects of reality that are still deeply hidden from our understanding.

Along the journey, humanity will also achieve its unfinished business and hopefully have no regrets about what could have been. It is here that we can start to link back up to the futures studies' literature. From Wendell Bell: There is a prima facie obligation of present generations to ensure that important business is not left unfinished.[22] We will not know what this unfinished business is until we encounter it on our epic journey. What important business this is goes unstated in both cases because, frankly, we cannot know what this business might be 1,000 years from now or even one million years from now. Even if we as a species achieve great things, such as terraforming a planet 100 light years from the earth by the year 10,000, there will most certainly be subsequent challenges and achievements in the offing. Richard Slaughter makes a similar point when he says that the "human project is unfinished."[23]

The journey is a meta-concept in that it is abstracted from everyday life in most ways. However, this does not mean that the journey and caring for future generations cannot also help meet the psychological needs of individuals. Both Bell and Slaughter acknowledge that caring for future generations can bring psychological benefits to current generations. Bell states that: The present generation's caring and sacrificing for future generations benefits not only future generations but also itself. Slaughter states that: Caring for future generations is a cultural force that is valuable now and for the foreseeable future.

Relating back to the literature in psychology, it should also be noted that self-actualization also encompasses other psychological needs, such as love and belongingness. In the broader case, we add the love received from and love given to past and future generations to the love received from family and friends. In Roger's terminology, this love can be unconditional.[24] We can also add the feeling of belongingness with all of humanity to the connection we feel in nature[25] and our socio-cultural contexts.[26] Lastly, there is much to be achieved with respect to the journey and each individual's achievements to further the journey that can be also be a positive source of psychological satisfaction.[27]

Thus, the journey is one toward humanity's self-actualization. The journey is deeply intertwined with our collective unconsciousness and completely consistent with larger psychological themes. The journey will be epic in many regards and will require heroism, too.

The journey: the biggest picture contract

John Rawls employed the concepts of *original position* and *veil of ignorance* in his seminal book, *A Theory of Justice*, to explore the qualities of a just society.[28] These concepts are the cornerstones to a thought experiment: what type of society would you design if you found yourself essentially at the beginning of time (original position) and without knowledge of when you would live or what your position in society would be (veil of ignorance)? In his thought experiment, Rawls allowed the participants to know certain fundamental interests they all have, plus general facts about psychology, economics, biology, and other social and natural sciences.[29] Rawls meticulously develops two major principles to guide the structure of a just society:

- First Principle – Each person is to have an equal right to the most extensive total system of equal basic liberties compatible with a similar system of liberty for all.
- Second Principle – Social and economic inequalities are to be arranged so that they are both:

  (1) to the greatest benefit of the least advantaged, consistent with the just savings principle, and

(2) attached to the offices and positions open to all under conditions of fair equality of opportunity.

The concepts of original position and veil of ignorance are also employed herein to explore justifications for protecting the equitable and sustainable journey of future generations through time and space. Let's play with these concepts with respect to the context of this book before returning to assess Rawl's principles.

Let's begin by reimagining the original position. With respect to the journey, we have some notions of when the journey may end, say one billion years from now unless we become extinct in the meantime. But when in time should we assume the journey begins? The answer could be 'now' for the sake of discussion purposes. But maybe we should set the discussion back about 10,000 years to the time just before the Agricultural Revolution or 300,000 years ago, when Homo sapiens appeared to have evolved?[30] Or maybe approximately 2.8 million years ago when the first species of Homo appear in the fossil record?[31] Or maybe 550 million years ago when animals first emerged on the earth?[32] Or 3.8 billion years ago when life first began on the earth? Or 4.5 billion years ago when the earth was formed? Or 13.8 billion years ago when the universe came into being after the Big Bang?

The choice of the time of the original position has a major bearing on how to manage the veil of ignorance. If we set the original position far enough back in time, then the very notion of Homo sapiens as participants in this discussion seems presumptuous, as we had not yet evolved and, if you believe Gould's premise, there would be no guarantee that we or any other intelligent species would ever evolve. By intelligence, I am referring to species that could conceive of this thought experiment and address the ethical implications of protecting the journey for future generations.

Let us say that life in spirit participated in a discussion in the original position. Far enough back in time, the veil of ignorance would have been quite heavy, only allowing that the participants would possess the spirit of life but would not even know their species, be they bacteria, dinosaurs, or humans. What would seem like a just contract amongst these life spirits? I argue that the just contract would be to imbue life with the will to live to ensure the journey of life – to live and evolve life to minimize the chance of the extinction of earth life. This is the most basic contract and most powerful contract imaginable. And so, it seems this contract is fulfilled as life on earth is absolutely driven to survive and diversify to strengthen its ability to survive. Additionally, as argued earlier, this will for the journey of life is hardwired into the human collective consciousness and concerns for future generations map well to satisfying our psychological needs.

Of course, we now know that to avoid extinction requires more than the shear will to survive, ecological diversity, and evolution. It also requires a consciousness to recognize systemic extinction risks and anticipate risks over very long periods of time. It requires engineered solutions to reduce risks and employ exquisitely sophisticated technologies that may also take millennia to develop and deploy. Now that we know this, it is appropriate to lighten the veil of ignorance to the current period of time to allow this knowledge to infuse the discussion. As the saying goes, today is the first day of the rest of humanity's life. However, the participants still do not know when in time they will exist, or whether they will be human, transhuman, or something else entirely.

This original position seems to be consistent with Rawl's thought experiment. However, instead of assuming a time frame that may be generational in scope, we are now asking participants to consider a one-billion-year time horizon. The participants are armed with knowledge about the nine categories of existential risks, Fermi's paradox, and the Drake equation. They are also armed with the realization that a strict utilitarian approach to public policy is wholly inadequate.[33]

It seems to me that the essence of Rawl's first principle might emerge in such a discussion. The discussants would not abandon the necessity for maintaining life's journey, but the discussion might be enhanced with the notion of the right to the opportunity to participate in the journey. This position is supported by beliefs that we should care about future generations because they have the right not to be harmed by actions taken today.[34] I believe that the discussants would consider it the duty of intelligent beings to make conscious decisions to protect the journey. Following Heilbroner,[35] we have a duty to care. This duty comes part and parcel with our being alive, independent, and conscious beings that are capable of planning and ethical reasoning. Duty does not allow the luxury of choice, as previously discussed.

Current generations fulfill their duties by meeting perpetual obligations that are needed to best protect the journey. These are discussed in much more depth in Chapter 6. Amongst these perpetual obligations are several that relate to equality, as contained in Rawl's first principle. Rawl's second principle is totally subsumed by one specific perpetual obligation, which requires current generations to bequeath sustainable societies to future generations.

Thus, in most ways, Rawl's two principles of a just society are quite consistent with the context of this book. The one area that is less consistent deals with Rawl's just saving principle. Here, he argues that it is justifiable to save for two generations, for one's children and grandchildren. After that, the contact to care about future generations seems to lessen, if not peter out completely. I believe this aspect of Rawl's work needs to be revised to be inclusive of all generations.

It also should be mentioned that the focus on journey from a Rawlsian perspective forcefully addresses three major criticisms leveled at the very plausibility of caring about future generations. One criticism posits that caring for future generations is impossible because obligations can only be made between living and identifiable beings.[36] The second one, usually put forth by economists, is that we cannot really care about future generations because we do not know what their preferences (i.e., utility functions) will be. The third criticism is that any efforts to regulate human behavior to, say, reduce the chances of catastrophic climate change, would reduce human freedom and therefore should be avoided.[37]

To address the third point first, protecting the journey in times of peril may reduce some freedom but it protects the ultimate freedom, the freedom to exist. This is a contractual duty that comes with existing and partaking of life. Freedom over many generations will not be curtailed if humanity is vigilant, anticipates risks, and is proactive.

Protecting the Spaceship Earth does not need recourse to arguments that we cannot know future generations' preferences. We assume that there are some guiding and universal preferences, as outlined earlier and in Chapter 6. Then everyday preferences can unfold as they will if a good job is done of maintaining options. Protecting the journey also does not need recourse to knowing the identities of future individuals.[38] The journey is protected for all future individuals, per agreement in the original position under the veil of ignorance. This is actually consistent with much of what happens already today. For example, urban planners develop comprehensive land use plans, and civil engineers design and construct massive urban infrastructures for the elements of the journey known as cities. People can freely come and go from the cities and it is fully expected that cities will outlive any specific inhabitants, much like the Ship of Theseus, where the ship, the container, exists as a functional entity even though, over time, every component is replaced.[39]

Why support future generations: obligations to earth life

Much has been written about humanity's ties to nature. For example, Komatsu has written that: Human beings have no eternal unchanging self-nature. We breathe in oxygen the plants breathe out. We are supported by that which is produced by nature. It is nature that blesses us with food, clothing, and dwellings.[40] This author takes a Buddhist perspective on how inextricably interwoven human lives are with nature. Humans cannot exist without the other earth life. Land should be loved and respected.[41] We should unconditionally care about nature because nature has intrinsic value.[42] We might even consider whether animals[43] or even trees have rights,[44] as have been granted to rivers in New Zealand and Ecuador.[45]

As important and as fascinating the topic of environmental ethics is, we need to remain on point for this chapter, which is why it is important to care about future generations of humans. The argument made in this section is that if one cares about the environment vis-à-vis the existence of other species of earth life into the very distant future, then one should also care about maintaining the existence of future generations of humans. Here is the argument.

The first point to be addressed is that some believe that humans ought to become extinct for the benefit of all other life on earth. Many argue that human behaviors have initiated a sixth massive species extinction[46] and are so disgusted with the negative impacts that humans are having on the planet that they argue that we deserve to go extinct. This is certainly the view of the Voluntary Human Extinction Movement.[47]

Let's put aside for the moment that humanity may very well become extinct for the many other reasons discussed in the previous two chapters. Let's also put aside the judgment made in the next chapter that humanity is by-and-large not currently meeting its obligations to future generations, which includes protecting the essence of nature and preventing the extinction of earth life. Let's focus on what is necessary for life on earth to avoid additional mass extinctions over the next several millions of years and to transcend the oblivion of our sun: *intelligence.*

Even though earth life has a strong genetic pre-disposition to survive,[48] this trait by itself is inadequate to ensure survival into the distant future. Sixty-five million years ago an asteroid smashed into the earth, caused rapid and catastrophic global climate change, and started a mass extinction event that ended the era of the dinosaur. The most extensive extinction event occurred roughly 250 million years ago, when approximately 96% of the species went extinct. A combination of climate change and a fatal decrease in the amount of oxygen in the atmosphere (which led to the extinction of most oxygen-breathing species) are cited as possible causes.[49] The 'non-intelligent' species alive at these times in the past had no capability to foresee impending doom nor to plan to prevent the catastrophes nor to save themselves. It was only by chance that earth life survived those two and three other massive extinction episodes. The situation is different now. Humans have the ability to plan, to act proactively, and to bring technology to the challenge.

Still, the question is whether it makes sense from the perspective of other earth life for humanity to survive or become extinct. A simple thought experiment suggests that humans are earth life's best bet. In this experiment there are three key factors: the probability that humans can avoid extinction and transcend the oblivion of the sun taking earth life with it; the probability that new intelligent life would re-evolve if humans became extinct; and the probability that a newly evolved intelligent species could avoid its own extinction and transcend oblivion, assuming there is enough time to do so.

The keys to this thought experiment are the second two components. Let's first address the possible reemergence of intelligent life on earth. First, if humanity were to become extinct, it is unlikely that similar intelligent life could re-evolve soon enough to lead earth life's effort to transcend the sun's oblivion. It took approximately 3.8 billion years for life on earth to evolve that has the potential intelligence to find homes in other solar systems and maybe even other galaxies. Any SCES that led to humanity's demise probably will also result in the extinction of most if not all other 'higher' life forms on the earth. Let's say that evolution is reset back to just before mammals and dinosaurs evolved 225 million years ago.[50] If intelligent life were to re-evolve in 225 million years, then that would be enough time to achieve the goal of transcending oblivion, but earth life would be vulnerable to additional mass extinctions and maybe total extinction during that very long time period.

Additionally, it is just a hypothesis that intelligent life would actually re-evolve in 225 million or even in another 3.8 billion years. Stephen Gould, a paleontologist and evolutionary biologist, believes that evolution of intelligent life (e.g., H. sapiens) on this planet was a random event and a rather unlikely event at that.[51] If the history of the earth were re-run with only a few initial parameters changed, it is very unlikely that intelligent life would have evolved at all. If humans and most animal species became extinct in the near term, could intelligent life re-evolve in time (a few billion years at most and maybe only over several hundred million years[52]) to save earth life from oblivion? Very possibly not.

Let's address the other two components simultaneously. The questions are whether humanity or a newly evolved intelligent species could avoid their own extinctions, help the earth avoid additional mass extinctions, and transcend the oblivion of the sun. The issue boils down to whether newly evolved intelligence species would be more, less, or equally likely to achieve these goals than the current dominant intelligent species, Homo sapiens. Essentially, we do not know whether newly evolved intelligent life forms would be better than humans or not. Given Fermi's Paradox, the fact that we have not detected other intelligent lifeforms in the university may suggest that the answer at best is equally likely. The fact that humanity has not yet gone extinct and it appears that most or all other intelligent species have disappeared from the universe suggests that it is very unlikely that a newly evolved intelligent species could do better than humanity, no matter how badly we believe humanity is doing currently.

To summarize, a newly evolved intelligent species might at best be equally able or more probably less able than Homo sapiens to protect earth life from future mass extinctions and the oblivion of the sun. However, it is not clear that intelligent life would evolve in time to achieve these goals or even re-evolve at all. One could argue that the probability of re-emergence in time could be quite low, say one in a billion. If this is the case, then earth life's best chances are with humanity and those who care about the rest of life on earth should then also have the motivation to care about future generations of humanity.

Conclusions

It is very plausible that in the future people will care more for future generations than they do now. Caring for future generations comports with deeply held values related to justice, virtue, and goodness. This chapter argues that we should care about humanity's self-actualizing journey through time and space. The notion of the journey is deeply resonant with human myth and psychology. Application of the concepts of original position and veil of ignorance suggests that the journey could also be a duty-bound contract that guides human decision-making over a very long period of time. It is also argued that humanity not only should protect its journey for itself but also for all other species of earth life as we are earth's most logical hope to transcend the oblivion of the sun.

Table 5.1 presents my subjective judgment about the extent to which humanity currently has the *propensity* to support each of the four reasons why it is important to care about future generations. I do not think many people have considered any reasons why we should care about future generations. I am using the word propensity to indicate my judgment about what percentage of individuals would support each reason after hearing the argument for each reason. As the table indicates, I believe that most individuals would agree that we should value caring for future generations. I also believe that the notion of the journey would resonate well. Thereafter, I think support for the other reasons would be lower in part because of practical, everyday living concerns.

Therefore, caring about future generations is a necessary component to the program set out in this book but in and of itself is not sufficient. Many other conditions are needed to transform caring into effective behavioral change. People need to know what to do to protect the journey and how well they are doing. The Perpetual Obligations presented in the next chapter and the metrics presented in Chapter 7 achieve these two goals. Efforts to protect the interests of future generations need institutional frameworks, which are addressed in Chapter 8. People's lives and livelihoods need to be enjoyable and sustainable. There has to be trust that pursuing a program to protect the journey of humanity through time and space can be accomplished while it is also possible to have a high quality of live and political and economic stability. These important topics are addressed in Chapter 9.

Table 5.1 Assessment of Support for Reasons for Caring About Future Generations

Reason to Care About Future Generations	*Propensity to Support*
Deeply Held Values	70%
Supporting the Journey	50%
Duty-Bound Contract	10%
Protect Earth Life	10%

The next chapter, then, focuses solely on presenting obligations or duties that current generations have to future generations. Since the refocus is on the journey, we are replacing the term obligations to future generations with the term *perpetual obligations.*

Notes

1 S. J. Gould. 1989. *Wonderful Life: The Burgess Shale and the Nature of History.* W.W. Norton, New York.
2 http://7genfoundation.org/7th-generation/
3 R. Heilbroner. 1995. *Visions of the Future: The Distant Past, Yesterday, Today and Tomorrow.* Oxford University Press, New York.
4 E. Partridge (Ed.). 1981. *Responsibilities to Future Generations: Environmental Ethics.* Prometheus Books. Buffalo, New York.
5 T. Mulgan. 2006. *Future People.* Oxford University Press, Oxford.
6 W. Bell. 1993. Why Should We Care About Future Fenerations? In H. Didsbury (Ed.), *The Years Ahead: Perils, Problems, and Promises.* World Future Society, Washington, DC, 25–41.
7 R. A. Slaughter. 1994. Why We Should Care for Future Generations Now. *Futures,* 26, 1077–1085.
8 C. Phillips. 2004. *Six Questions of Socrates.* W.W. Norton, New York.
9 Ibid., 289.
10 http://artsites.ucsc.edu/gdead/agdl/truckin.html
11 www.inverse.com/article/34170-tardigrade-water-bear-asteroid-supernova
12 https://en.wikipedia.org/wiki/Tardigrade; www.wired.com/story/a-crashed-israeli-lunar-lander-spilled-tardigrades-on-the-moon/
13 C. Jung. 1959. *Collected Works, Vol. 9.1. The Concept of the Collective Unconscious (1936).* J. Jacobi, Complex/Archetypes/Symbol in the Psychology of C. G. Jung, 42. Pantheon Books, New York.
14 https://frithluton.com/articles/heroic-journey-jungian-perspective/
15 C. Jung. 1959, ibid.
16 K. Armstrong. 2005. *A Short History of Myth.* Canongate Books, London.
17 J. Campbell. 1988. *The Power of Myth.* Anchor Books, New York.
18 J. Campbell. 1968. *A Hero with a Thousand Faces.* Princeton University Press, Princeton, NJ.
19 A. Maslow. 1970. *Motivation and Personality.* Harper and Row, New York.
20 https://en.wikipedia.org/wiki/Self-actualization
21 www.simplypsychology.org/maslow.html
22 W. Bell. 1993, ibid.
23 R. A. Slaughter. 1994, ibid.
24 C. Rogers. 1959. A Theory of Therapy, Personality, and Interpersonal Relationships, As Developed in the Client Centered Framework. In S. Koch (Ed.), *Psychology: A Study of Science,* vol. 3. McGraw-Hill, New York, 184–256.
25 L. Binswanger. 1963. *Being-in-the-World: Selected Papers of Ludwig Binswanger.* Basic Books, New York.
26 E. Goffman. 1956. *The Presentation of Self in Everyday Life.* Doubleday, New York.
27 D. McClelland. 1961. *The Achieving Society.* Van Nostrand, Princeton, NJ.
28 J. Rawls. 1971. *A Theory of Justice.* Harvard University Press, Cambridge, MA, 302.
29 https://plato.stanford.edu/entries/original-position/
30 https://en.wikipedia.org/wiki/Homo_sapiens
31 https://en.wikipedia.org/wiki/Human_evolution
32 https://sci.waikato.ac.nz/evolution/AnimalEvolution.shtml

33 D. Feldman. 1995. *Water Resources Management.* Johns Hopkins University Press, Baltimore, MD.
34 J. Brannmark. 2016. Future Generations as Rightholders. *Critical Review of International Social and Political Philosophy*, 19, 6, 680–698.
35 R. Heilbroner. 1975. What Has Posterity Ever Done for Me? *New York Times*, January 19.
36 D. Parfit. 1984. *Reasons and Persons.* Clarendon Press, New York.
37 N. Oreskes and E. Conway. 2010. *Merchants of Doubt.* Bloomsbury, New York.
38 J. Reiman. 2007. Being Fair to Future People: The Non-Identity Problem in the Original Position. *Philosophy & Public Affairs*, 35, 1, 69–92.
39 www.amazon.com/Ship-Theseus-J-Abrams/dp/0316201642
40 As quoted by S. Shimizu. 1992. *How Long Can We Survive, in Voices from Kyoto Forum in Earth Summit Times.* Kyoto Forum, Osaka, Japan, 146.
41 A. Leopold. 1949. *A Sand County Almanac.* Oxford University Press, Oxford.
42 A. Næss. 1973. The Shallow and the Deep, Long-Range Ecology Movement. *Inquiry*, 16, reprinted in Sessions 1995, 151–155.
43 P. Singer. 1975. *Animal Liberation.* Random House, New York.
44 C. D. Stone. 1972. Should Trees Have Standing? *Southern California Law Review*, 45, 450–501.
45 https://theconversation.com/when-a-river-is-a-person-from-ecuador-to-new-zealand-nature-gets-its-day-in-court-79278
46 R. Leakey and R. Lewin. 1995. *The Sixth Extinction: Patterns of Life and the Future of Humankind*, Doubleday, New York.
47 https://en.wikipedia.org/wiki/Voluntary_Human_Extinction_Movement
48 R. Dawkings. 1989. *The Selfish Gene.* Oxford University Press, Oxford, England.
49 P. Ward. 2004. *Gorgon: Paleontology, Obsession, and the Greatest Catastrophe in Earth's History.* Viking Press, New York.
50 https://evolution.berkeley.edu/evolibrary/article/0_0_0/evotext_13
51 S. J. Gould. 1996. *Full House: The Spread of Excellence from Plato to Darwin.* Harmony Books, New York.
52 P. Ward and D. Brownlee. 2002. *The Life and Death of Planet Earth: How the New Science of Astrobiology Charts the Ultimate Fate of Our World.* Times Books, New York.

6 Twelve perpetual obligations

The previous chapter presented four arguments for why we ought to care about future generations and maintaining humanity's journey of self-actualization into the distant future. It is argued that devotion to these goals is ethical, can meet fundamental psychological needs, is consistent with rational discourse on this topic, and is harmonious with our obligations to earth life. As noted in the previous chapter, I believe that every human of every generation has a duty to help perpetuate the journey.

The purpose of this chapter is to propose a new set of twelve perpetual obligations that, if met, will help ensure that humanity will maintain its journey through time and space. Perpetual obligations are actionable goals that need to be addressed by contemporary public policies. The first section of this chapter summarizes previously published statements about obligations to future generations. This sets the stage for this new set of perpetual obligations which can be viewed as: updates to previously published obligations to future generations; a more comprehensive set of obligations; and a set designed to link to public policy. Indeed, accompanying the definition of each perpetual obligation is a short discussion about how we can measure whether the obligation is being met or not. The chapter includes a discussion about how the dozen perpetual obligations map to the literature and a scorecard that captures my judgments on whether humanity is currently meeting each perpetual obligation. An extended discussion about how to set an ethical threshold related to an acceptable probability of human extinction concludes the chapter.

Literature review

As has already been noted, futurists, philosophers, and others have written extensively on the topic of future generations.[1] In Chapter 5, we introduced frameworks proposed by Bell and Slaughter that deal with both why we should care about future generations and what those obligations should be.[2] We revisit these statements of obligations here and add some additional statements. These are consolidated in Table 6.1 and also in Exhibit 6.1.

Table 6.1 Statements on Obligations to Future Generations

Source	*Statement*
Bell	Regarding the natural resources of the earth, present generations have no right to use to the point of depletion or to poison what they did not create
	Past generations left many of the public goods that they created not only to the present generation but to future generations as well
	There is a *prima facie* obligation of present generations to ensure that important business is not left unfinished
Slaughter	The human project is unfinished
	We are partly responsible for the dangers to their well-being
	The global commons has been compromised by human activity and restorative actions are necessary
McClean; Schrader-Frechette	Fairness obligation with respect to risk
Weiss; Golding; Tonn; Balazs and Gaspar; Gilroy	Maintaining options obligation
Tough	Peace and security
	Environment
	Children and learning

In my opinion, three of Bell's and three of Slaughter's statements put forth obligations to future generations. Bell's pertain to sustainability, protection of humanity's hard-won knowledge base, and the no-regrets obligation previously mentioned. Similarly, Slaughter's statements pertain to the no-regrets obligation and sustainability issues. Added to the mix is a statement about risk to the well-being of future generations, which is generically stated.

Not fully developed by these two authors are two additional viewpoints on obligations to future generations that are eminently amenable to operationalization from a policy perspective. First, the fairness obligation concerns not imposing risks on future generations that current generations would also not accept. For example, MacLean et al.'s neutrality criterion states that "levels of risk to which future generations will be subjected will be no greater than those of present persons."[3] Risks can include premature death from environmental or other preventable catastrophes.[4] Fairness also has an element of consent. According to Schrader-Frechette,[5] "Until or unless a risk imposition receives the consent of those who are its potential victims, it cannot be justified."

Second, Weiss,[6] Tonn,[7] Golding,[8] Balazs and Gaspar,[9] and others argue for a "maintaining options" obligation, meaning that decisions made by current generations should not restrict possible futures that could be pursued by future generations. Weiss's Principle of Conservation of Options holds that: "Each generation should conserve the diversity of the natural and cultural resource base so that it does not unduly restrict options available to future generations." Additionally, the maintaining options obligation entails gifting to our posterity future worlds that are as free of man-made constraints as possible. In other words, there is a need to prevent environmental and other catastrophes "that would restrict the future of the human race by cutting off certain possible futures."[10] By cutting off many futures, the ability of future societies to grow and mature is reduced[11] as is the freedom for people to "reason about means and ends and evaluate preferences, to match desires and beliefs and then act."[12]

Tough states that future generations need these types of things from current generations:

> Future generations need equal opportunity (a legacy as beneficial as ours was), our caring about their well-being, and attention to their needs in our legislatures and parliaments. Their particular needs are for us to focus on peace and security, the environment, the worst risks of all, governance, the knowledge base, children and learning (p. 1041).[13]

Tough adds several additional important notions to the mosaic, including peace and security and a focus on children and learning.

In 1997, the United Nations Education, Science and Culture Organization (UNESCO) approved a Declaration of Obligations to Future Generations.[14] The twelve articles of this declaration are presented in full in Exhibit 6.1 and represent the most comprehensive set of statements found in the literature. For example, Article 2 directly addresses the maintaining options obligation. Article 3 introduces the notion of perpetuity, which I interpret as an obligation to prevent humans from becoming extinct. Article 4 can be interpreted to mean we also need to ensure that earth life does not become extinct. Article 6 addresses the essence of both nature and the human species that we have already touched upon. Articles 7 and 8 stress that current generations have an obligation to protect humanity's cultural heritage. Lastly, articles 9 and 10 mirror Tough's concerns about peace and education.

Exhibit 6.1 UNESCO Declaration of Obligations to Future Generations

Article 1 – Needs and interests of future generations

The present generations have the responsibility of ensuring that the needs and interests of present and future generations are fully safeguarded.

Article 2 – Freedom of choice

It is important to make every effort to ensure, with due regard to human rights and fundamental freedoms, that future as well as present generations enjoy full freedom of choice as to their political, economic and social systems and are able to preserve their cultural and religious diversity.

Article 3 – Maintenance and perpetuation of humankind

The present generations should strive to ensure the maintenance and perpetuation of humankind with due respect for the dignity of the human person. Consequently, the nature and form of human life must not be undermined in any way whatsoever.

Article 4 – Preservation of life on earth

The present generations have the responsibility to bequeath to future generations an Earth which will not one day be irreversibly damaged by human activity. Each generation inheriting the Earth temporarily should take care to use natural resources reasonably and ensure that life is not prejudiced by harmful modifications of the ecosystems and that scientific and technological progress in all fields does not harm life on Earth.

Article 5 – Protection of the environment

(1) In order to ensure that future generations benefit from the richness of the earth's ecosystems, the present generations should strive for sustainable development and preserve living conditions, particularly the quality and integrity of the environment.
(2) The present generations should ensure that future generations are not exposed to pollution which may endanger their health or their existence itself.
(3) The present generations should preserve for future generations natural resources necessary for sustaining human life and for its development.
(4) The present generations should take into account possible consequences for future generations of major projects before these are carried out.

Article 6 – Human genome and biodiversity

The human genome, in full respect of the dignity of the human person and human rights, must be protected and biodiversity safeguarded.

Scientific and technological progress should not in any way impair or compromise the preservation of the human and other species.

Article 7 – Cultural diversity and cultural heritage

With due respect for human rights and fundamental freedoms, the present generations should take care to preserve the cultural diversity of humankind. The present generations have the responsibility to identify, protect, and safeguard the tangible and intangible cultural heritage and to transmit this common heritage to future generations.

Article 8 – Common heritage of humankind

The present generations may use the common heritage of humankind, as defined in international law, provided that this does not entail compromising it irreversibly.

Article 9 – Peace

(1) The present generations should ensure that both they and future generations learn to live together in peace, security, respect for international law, human rights, and fundamental freedoms.
(2) The present generations should spare future generations the scourge of war. To that end, they should avoid exposing future generations to the harmful consequences of armed conflicts as well as all other forms of aggression and use of weapons, contrary to humanitarian principles.

Article 10 – Development and education

(1) The present generations should ensure the conditions of equitable, sustainable, and universal socio-economic development of future generations, both in its individual and collective dimensions, in particular through a fair and prudent use of available resources for the purpose of combating poverty.
(2) Education is an important instrument for the development of human persons and societies. It should be used to foster peace, justice, understanding, tolerance, and equality for the benefit of present and future generations.

Article 11 – Non-discrimination

The present generations should refrain from taking any action or measure which would have the effect of leading to or perpetuating any form of discrimination for future generations.

Article 12 – Implementation

(1) States, the United Nations system, other intergovernmental and non-governmental organizations, individuals, public and private bodies should assume their full responsibilities in promoting, in particular through education, training, and information, respect for the ideals laid down in this Declaration, and encourage by all appropriate means their full recognition and effective application.

(2) In view of UNESCO's ethical mission, the Organization is requested to disseminate the present Declaration as widely as possible, and to undertake all necessary steps in its fields of competence to raise public awareness concerning the ideals enshrined therein.

Perpetual obligations[15]

This section presents a list of twelve perpetual obligations. Meeting the perpetual obligations represents one answer to Ord's question: *What can humanity do right now that will make the biggest difference over the next billion years?*[16] The term 'perpetual obligation' was introduced in the previous chapter to signal the shift from the term 'obligations to future generations', which, in my opinion, unwisely allows current generations to choose whether they have any obligations to future generations, to obligations that all current generations must meet as a matter of course over all time. The first eight obligations focus on preventing the emergence of horrendous potential futures. If these risks are not present (i.e., the obligations are fully being met), then no actions are needed. The remaining four obligations are proactive in that humanity needs to take every year into the distant future to help ensure that disastrous futures can be avoided or ameliorated. Perpetual obligations are defined such that they are measurable and actionable. Thus, accompanying the definitions are proposed metrics to measure how well each perpetual obligation is being met.

Perpetual Obligation #1. Prevent the risk of human extinction from exceeding the ethical threshold over an appropriate time horizon

All generations have a perpetual obligation to prevent human extinction. From our vantage point in the first half of the twenty-first century, I argue that the acceptable threshold for the probability of human extinction is

10^{-20} (the approach used to develop this estimate is found at the end of this chapter). The assessment metric, then, is the probability of human extinction. It is difficult to rigorously estimate the probability of human extinction because we have no historical data to go by. We have seen in Chapter 3 that various authors and researchers have provided their own subjective estimates, presumably based in part upon their own idiosyncratic synthesis of the risks humanity faces and some understanding of probability theory. While these estimates are quite high compared to the 10^{-20} threshold, they are not based on rigorous and replicable analytical methods. Also presented at the end of this chapter are some estimates of the probabilities of extinction-level events, which are also high respect to the 10^{-20} threshold. However, as suggested in Chapter 4, these probabilities do not consider human behavior.

This all said, let me propose an approach that can be taken in the near term and improved or even replaced in the mid-to-long term to estimate the probability of human extinction. Central to this approach are the singular chain of event scenarios (SCES) introduced in Chapter 4. The approach entails developing a very large number of reference SCES, say several thousand. Their development could be crowdsourced across the globe. Each would be vetted for plausibility. An AI-equipped support system could be developed to interactively help SCES writers and to help vet the results as well.[17] Also, each would carry a 'pedigree' of metadata that describes key aspects of each scenario's genesis and plausibility. For example, the pedigree could indicate which category of extinction event was the main triggering event, when in time the scenario would end (e.g., 100 years, 1000 years), if the scenario is representative of a larger group of SCES (i.e., just one version) or is truly singular, and if its likelihood of occurring appears to be greater than 10^{-20}. An institution that could be given responsibility for developing and maintaining the scenario repository is addressed in Chapter 8. In fact, this institution would shepherd a set of themed scenarios needed to assess whether humanity is meeting its perpetual obligations. These are described in Exhibit 6.2.

A worldwide Delphi panel of experts would through an iterative process use the reference SCES and other important data (e.g., from global climate models, forecasts about the availability of energy supplies) to estimate the probability of human extinction over a 1000-year time frame. It is also necessary to have some idea about how many potential paths into the future humanity could take during these time frames so that the panelists could estimate the probability of human extinction as the ratio of the number of SCES over the total number of potential paths into the future. An international organization is proposed in Chapter 8 that would have responsibility for working with other organizations around the world to convene a reputable and trusted panel of experts.

Perpetual Obligation #2. Prevent the risk of total extinction of all life on earth from exceeding the ethical threshold over an appropriate time horizon

This perpetual obligation has been addressed previously, most directly in Chapter 5. This obligation relates directly to current and future generations of humans since human life could not exist without the existence of other species of earth life.[18] It acknowledges the fact that human biology is tightly coupled with the global ecosystem[19] and our bodies are even infused with non-human species.[20] This obligation can also be viewed as a fundamental obligation to earth life overall.[21] Thus, one can argue that the threshold for preventing the extinction of all life on earth needs to be even stricter than the risk threshold with respect to human extinction. It is argued elsewhere that this threshold should be 10^{-23} over a 1000-year time horizon.[22]

SCES are also needed to help assess whether the 10^{-23} threshold is being met. It is recommended that a subset of human extinction SCES be expanded into Earth-life Extinction Scenarios (ELExS) to explore the extinction of earth life as well. After all, each earth life extinction scenario would necessarily also encompass our own extinction. Of course, not every SCES will then move on to posit the extinction of all earth life.

Perpetual Obligation #3. Bequeath sustainable societies

Sustainable societies have two essential characteristics; they are humane, and they are also able to make the decisions and carry out programs over the long term to perpetuate humanity's journey through time and space. I believe that these two characteristics are inextricably linked.

It is not enough for humanity to avoid extinction. Life also must be worth living in a fundamental sense. Thus, current generations have an obligation to bequeath to future generations strong institutions,[23] traditions,[24] and societies that are just and peaceful, fundamentally stable and civilized, essentially following the thoughts of Tough and Rawls. The societies could be imbued with very strong democratic processes, as advocated by Barber.[25] Certainly, these societies need to advocate and protect human rights and promote cultural diversity and heritage, as highlighted by UNESCO's declaration of obligations to future generations.

Exhibit 6.2 Themed Scenarios

Development and assessment of perpetual obligation metrics can be supported by several types of scenarios. We have encountered one type of scenario already, Singular Chains of Events Scenarios (SCES), introduced in Chapter 4. The purpose of each SCES is to describe how a very specific path over time ultimately leads to the extinction of humanity. As such, this scenario has a theme and a predefined end point. These scenarios are needed to assess whether humanity is meeting perpetual obligation #1, which is related to human extinction and indirectly support the metrics for perpetual obligations #3, #4, #10, and #12. Three such scenarios are presented herein.

In this chapter, we need several additional types of themed scenarios to help estimate perpetual obligations metrics:

- Earth-life Extinction Scenarios (ELExS) – These scenarios describe paths over time where all earth life becomes extinct, including humans. These scenarios directly support the metric for perpetual obligation #2.
- Global Catastrophe Scenarios (GCS) – These scenarios describe paths over time that result in major loss of life and collapse of civilization, but do not ultimately result in human extinction. These scenarios directly support the metrics for perpetual obligation #4 and indirectly perpetual obligations #3, #8, #10, and #12.
- Muddling Through Scenarios (MTS) – These scenarios describe paths over time where humanity never quite meets its perpetual obligations but muddles along year after year, century after century, never quite achieving its potential while avoiding global catastrophes. These scenarios support the metrics for perpetual obligations #3, #4, #5, #6, #7, #8, #10, #11, and #12.
- Muddling Through Scenarios – High Emissions (MTS–HE) – In these scenarios, muddling through is unsuccessful in reducing involuntary environmental risks, while still avoiding global catastrophes. These scenarios are specifically designed to support the metrics for perpetual obligation #5.
- Transformative Scenarios (TS) – The concept of socio-cultural singularity was introduced as part of the serendipitous journey through future time presented in Chapter 2. The Singularity describes a future time where, in order for humanity to survive and flourish, almost every aspect of society has significantly and synergistically transformed. This topic is addressed thoroughly in Chapter 9, which also includes a TS at the end.

As noted, each instantiation of each type of scenario needs to describe paths over time. The figure below presents generic examples of each type of scenario. As shown, the SCES end when human population hits zero. The scenario starts in a state of concern and then worsens (from dots to light grey to dark grey to dashes) over time. In contrast, the MTS bumps along in a continual state of concern (dots) but without major changes in human population. The canonical TS passes through the singularity (T) at some point in the future. One may assume that TS allows for higher human populations, but that is not a necessary stipulation. It should also be noted that the SCES and GCS can terminate at any point in time within the designated planning horizon, which in this case is indicated to be 1000 years.

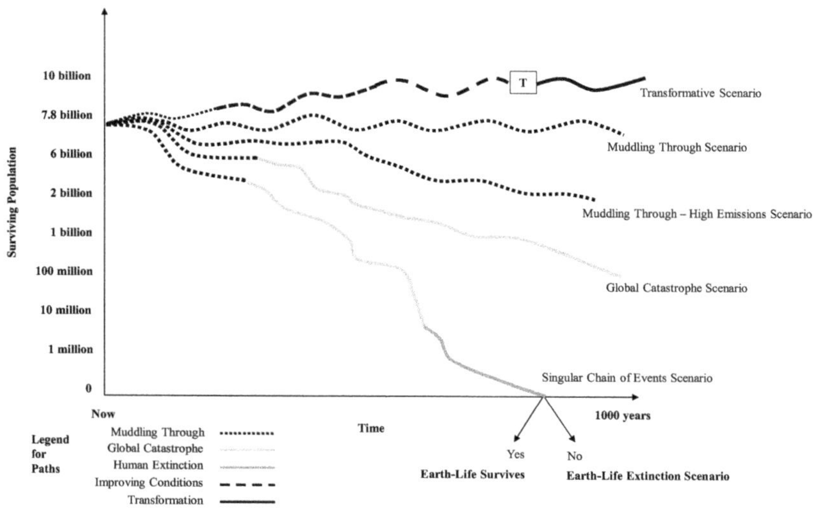

A final couple of points are these. There are no limits to the number of any type of scenario. Each scenario must have internal validity and be plausible. Each scenario will need to be vetted by a third party to certify validity and plausibility (see Chapter 8 for a recommendation for this third party). How to certify validity and plausibility for potentially thousands of scenarios are open research questions.

This leads to the second major aspect that the sustainable societies need to be capable decision makers. Societies need to be able to make timely and effective decisions to impede emerging CEs. These types of societies would be less roiled by internecine conflicts that would draw focus and energy from futures-decision-making. These types of societies would also be more likely to derive benefits from meeting their perpetual obligations.

For symmetry, I propose two approaches to assessing whether we are meeting this obligation. The first is to develop an index that measures sustainable societies. As a start, we could draw upon existing indices that measure freedom, democracy, and human rights. These include Freedom House's political rights and civil liberties indices[26]; Human Freedom

Index[27]; World Index of Moral Freedom[28]; MaxRange[29]; and Democracy Index.[30] The Human Freedom Index is composed of 76 indicators, which are aggregated into an overall index score that runs from 0 to 10, with 10 being the highest score possible. Ideally, the goal would be for all countries to score 10 (in 2017, the average score across 162 countries was 6.89). An essential question with respect to this and the other indices is what is the minimal score that each country would need to exceed in order to meet this perpetual obligation.

As a second approach, we can use all of the scenarios presented in Exhibit 6.2 to assess our abilities to make sustainable decisions. Imbedded within many if not all SCES will be decision nodes, or places, in the chain of events where decisions could have been made and were not made due to deficient and/or anti-sustainable decision-making processes. Certainly, deficient decision-making is at the core of many SCES and ELExS but is also intrinsic to the Muddling Through Scenarios and Global Catastrophe Scenarios. An important part of the scenarios' vetting process will be to use human knowledge about our own decision-making processes. The scenarios will be vetted within what is known and anticipatable about human decision-making, not with how we would prefer decisions to be made. This would be the case unless there is evidence that we have actually improved or could be anticipated to improve our decision-making processes. If a sizable proportion of scenarios is animated by deficient decision-making, then this would also be an indication that this perpetual obligation is not being met. Again, additional work is needed to ascertain what proportion of deficient decision-making scenarios over all scenarios would indicate that this perpetual obligation is not being met.

Perpetual Obligation #4. Bequeath sustainable systems of production

A common distinction made in the literature about sustainability is also being made here: that sustainability can be understood to have social, economic, and environmental aspects.[31] This specific obligation focuses on the use of resources needed to produce goods and services provided by global economies. Thus, herein sustainability is interpreted in this sense: can the magnitudes of resources anticipated being consumed in the global economy be sustained over the next 1000 years? Here I am thinking about energy, water, soils, and virgin materials. If the answer is no, then corrective actions are required. The more unsustainable the prospects appear, the more work current generations must do to rectify the situation. At the limit, a wholly unsustainable economic system could result in an existential risk to humanity.

Sustainability also pertains to the ability of the earth's systems to absorb without harm degradations caused by production. Are planetary boundaries with respect to systems such as ocean acidification or stratospheric ozone depletion being crossed?[32] If so, then humanity is not meeting this perpetual obligation.

Thus, metrics include shortages of key production inputs and crossed planetary boundaries that could lead to global ecosystem failure. It is suggested that scenarios and models be used to assess whether this obligation is being met. The SCES could be used but it would also be useful to have scenarios that describe unsustainable systems of production that also do not lead to human extinction, such as the Muddling Through Scenarios presented in Exhibit 6.2. Numerous models are developed and tended to by research groups across the world that focus on global environmental systems, energy production, water resources, agriculture, etc.[33]

Perpetual Obligation #5. Continuously work to reduce the risk of death from involuntary environmental risks that are greater than the ethical standard of 10^{-6}

Environmental regulations developed around the world are often designed to reduce the probability of an individual dying from a specific involuntary environmental risk to be below 10^{-6} on an annual basis.[34] This obligation is clearly subsidiary to the obligation to prevent human extinction and in fact can be understood to merge into this obligation if the accumulation of involuntary environmental risks threatens human extinction. This ethical threshold is a separate obligation in order to focus explicit attention on pollution, exposures to toxic substances, and environmental justice and to clearly indicate that sustainable economies also need to be clean economies.

Perpetual Obligation #6. Preserve the essence of nature

Up to this point, the list of obligations can be seen to have roots in decades of thinking about obligations to future generations, social and environmental justice, and sustainability. This perpetual obligation, to preserve the essence of nature, is motivated by more recent developments in, and concerns stemming from, new genetic engineering technologies. Without a doubt, humans have impacted the genetics of species for thousands of years. We have appropriated most of nature for our own needs.[35] We have altered crops and food and fiber-producing animals through selective reproduction and domestication. In recent decades, we have genetically modified numerous species for our own purposes (e.g., rice, alfalfa, soybean, and more). The potential for humans to change nature has greatly increased though the discovery and development of CRISPR technology.[36] This relatively inexpensive technology allows us to drive human-designed genes into the DNA of organisms in a way that ensures that these genes will be passed down to all succeeding generations. For example, as discussed in Chapter 4, researchers have discussed driving genes into mosquitoes to prevent malaria, into Asian carp to ensure that only males will be propagated (and thereby ensuring their elimination as an invasive species), and into American chestnut trees to help them withstand the chestnut blight.[37]

The concern expressed by this obligation is that an accumulation of modifications to a broad spectrum of organisms could cross an ethical threshold at which something essential about earth life is irreversibly lost to the detriment of all future life on earth. Earth life is characterized by diversity and evolution. Diversity provides stability and adaptability to threats to ecosystem processes. Evolution produces diversity within the framework of natural experimentation. We need to respect and protect diversity and evolution even if this means restraining and even refraining from manipulating nature to meet our interests in the short term. A world full of organisms designed to meet the needs of humans is unlikely to be a healthy world, literally and aesthetically. If manipulation of nature becomes rampant and unintended consequences prove devastating, then this obligation merges with the obligation to prevent the extinction of life on earth.

There is much debate about what is natural versus what is not natural. Humans have been shaping the natural world for thousands of years, maybe even tens of thousands of years if we note the unnatural extinction of species by the hands of our ancestors.[38] Thus, it is hard to propose a baseline that describes what nature is and, therefore, what its essence might be. Component parts of this metric could include the amount of the earth that is free from human appropriation (land, water, air) where native species can exist and evolve; the ability of nature to evolve at normal rates; the extinction of species that does not exceed normal rates; and the approximate percentage of non-domesticated species that are free from the influences of human genetic engineering. What the acceptable percentage of non-domesticated species with human-caused genetic modifications should be considered very seriously. If it is zero, then movements to use CRISPR to modify organisms to wipe out invasive species and protect indigenous species from invasive species would need to be banned immediately worldwide.

Perpetual Obligation #7. Preserve the essence of the human species

Advances in technology also motivate this obligation. Because of the evolution of technology, one can assume that there will be growing potentials to change, alter, and redesign ourselves to become transhuman.[39] Humans are already becoming increasingly cyborgian through technologies as diverse as knee and hip replacements to heart pacemakers and hearing aids. Technology is also advancing to overcome blindness and paralysis and to allow humans to mentally control artificial limbs. Future cognitive and genetic enhancements could greatly increase intelligence, memory, and life spans.[40]

These trends merely represent what is happening toward the beginning of the twenty-first century. One can only imagine what technologies of the twenty-second, twenty-third, and the succeeding centuries may offer. Thus, one must ask this question: at what point do we cease being human and become something else, something so essentially different that current humans have no obligations to this new life form? Or, even more starkly, do we have obligations to future generations that may be artificial, that is, robotic or virtual? In the extreme, one could argue that the emergence of these forms of artificial life could pose an existential risk to humans. It is important for current generations to work to preserve the essence of the human species.

Defining a baseline human is not straightforward. Different cultures and different generations may have different levels of comfort and familiarity with technologic and genetic enhancements. My own opinion is that uploaded human personalities that have no biological components are not human. On the other hand, I consider myself still human though I have benefited from vaccinations, antibiotics (starting with penicillin), vision correction, and the wonders of modern surgery. In any case, the range between a human devoid of any alteration and an uploaded personality is enormous and deserves careful consideration by philosophers, theologians, and all concerned humans.

Perpetual Obligation #8. Maintain options for future generations

This obligation is taken directly from the work of Brown and others cited earlier. Its value in this list is to make explicit that the evolution of humanity encompasses infinite potential futures and this freedom is to be explicitly valued and protected. Of course, this obligation is best met when the risks addressed by the first seven obligations are minimal or nonexistent. In the best of times, societies do not have to limit their options by allocating resources or regulating behaviors to meet obligations to future generations to prevent catastrophes. Of course, freedoms for future generations can be threatened in other ways. For example, restrictions against creativity and innovation can stifle the development of new paths for humanity to follow. Repression and the imposition of a limited set of prescribed behaviors is even more damaging. Extreme social controls in the absence of existential risks can threaten the ability for current generations to bequeath sustainable societies, and pathologically, actually increase existential risks.

This obligation straddles a tough issue for futures studies: not knowing the preferences of future generations versus ensuring that future generations can exist. The intent of the first seven perpetual obligations is to protect the inhabitability of earth's 'system', which includes foundational assumptions that future generations will prefer infinitely sustainable systems with respect to energy, food, water, etc. and will not prefer to suffer involuntary environmental risks.

We need to be open-minded about what maintaining options means. It basically means the freedom of action within the bounds of perpetual obligations. Future generations could decide they still love consumerism and capitalism. This is fine as long as the means of production are sustainable and resources are still devoted to managing extinction risks, and humanity is still working to protect the essence of the human species and nature.

Assessing whether this perpetual obligation is being met and anticipating whether it will be met in the future can be accomplished in a couple of manners. We can assess the range of laws, regulations, programs, and policies that are in place to meet perpetual obligations and are also in place for socio-cultural reasons. We can also refer to the various indices discussed earlier with respect to freedom. With respect to the latter, we can make use of all of the scenarios described earlier to assess how and to what extent options may be constrained along many, many potential paths into the future.

Perpetual Obligation #9. Preserve knowledge gained by previous and current generations for the use of future generations

The ninth perpetual obligation is for current generations to preserve knowledge gained by humanity over time for use by future generations. Knowledge can take many forms, from how to build a radio to the general theory of relativity, to archiving examples of extinct species and languages to successful and unsuccessful medical treatments. One can only speculate how much the loss of knowledge from the destruction the library in Alexandria, Egypt in 48 BC cost humanity over succeeding millennia. Or how the worldwide loss over time of parchments, scrolls, manuscripts, and early books and letters, not to mention knowledge transmitted through oral traditions, diminished the humanities and science.

Humanity must constantly work to meet this obligation as knowledge continues to grow and continuously needs to be preserved. The digitization of knowledge in recent years has made the exponential growth in knowledge available to billions across the globe, but the existence of the 'cloud' does not necessarily satisfy this obligation. In fact, the cloud may make humanity complacent with respect to this obligation because the cloud could disappear. Today's archivists can provide estimates of how much of our current knowledge is in digital form and how much is not. A major challenge will be to capture knowledge that is indigenous and/or analogue in nature. I imagine that efforts to document threatened indigenous languages could be expanded to include all forms of cultural knowledge.

The archival solutions need to be fail-safe yet allow the information to be easily accessible, such as through the WEMS. The metric for this perpetual obligation is the ratio of human knowledge that is archived in a fail-safe environment versus all human knowledge. Resources devoted to meeting this obligation may increase during periods where archives are transitioning from one set of solutions to new, better solutions or when events pose unexpected risks to the archives.

Perpetual Obligation #10. Actively support the generation of new knowledge needed to support the survival of earth life into the distant future

New knowledge needs to be continuously gained to help humanity and earth life survive into the distant future. This knowledge needs to serve the meeting of the other eleven perpetual obligations. A cursory review of research and development expenditures made by today's societies indicates funds are mainly allocated for defense, health (e.g., life span extension), and the well-being and enjoyment of current generations.[41] The preponderance of this research focuses on the natural sciences and on technology design, development, and deployment. Even within this realm, relatively few dollars are being spent to reduce risks to humanity *in toto* or even to reduce involuntary environmental risks.

It should be noted that new knowledge is not only restricted to the realm of the natural sciences and technology. We also need to learn how to better organize ourselves politically and economically. Humanity needs to develop peaceful and ethical means to experiment with new political and macroeconomic frameworks. These and other ideas are addressed in more depth in Chapter 9.

Compared to other perpetual obligations, the metric for this obligation is straightforward: how much of the world's resources are being spent with respect to this obligation. This amount could be stated in terms of percentage of world gross domestic product. The major question, then, is what should the percentage be? The answer could be conditional upon the risks humanity is facing and is anticipated to face. As a matter of course, humanity could be advised to reinvest at least 5% of its economic resources to maintaining its existence, more when new knowledge is absolutely required to deal with anticipated challenges.

Perpetual Obligation #11. Minimize regrets that humanity might have over not finishing its important business

Minimizing regrets that humanity might have over not finishing its important business (in the words of Wendell Bell[42]) is best satisfied when humanity is following its enlightened interests and working toward grand dreams, instead of remaining focused on the individual struggle to survive. This obligation stresses humanity over individuals. To meet this obligation, humanity needs grand dreams to pursue and a large dose of humility. Grand dreams can be pursued within numerous realms of human endeavor, from space exploration to fundamental science, from the arts to the humanities. We need to be humble enough to admit that humanity probably does not know what its important business is at this time. We need to be open-minded enough to recognize what this business might be and to be able to take advantage of opportunities to achieve humanity's grand dreams. It could be that to take advantage of a once in a million year's opportunity, humanity would have to band together to devote a very substantial amount of resources to a very special project.

The development of WEMS could represent one such dream. Another idea for a special project that humanity could undertake right now is the development of a global electricity grid. The idea, first proposed by Buckminster Fuller,[43] is to link the world together into one grid. This idea is appealing for several reasons. It could bring electricity to underdeveloped regions. It could hasten the electrification of our homes, businesses, and transportation technologies, which, if powered by renewable, inexhaustible, and nuclear resources, could greatly reduce the emissions of greenhouse gases. To achieve this vision, the global grid would also provide links to prodigious amounts of renewable resources that are located in sparsely populated regions and maybe even in space. This idea is also appealing because it would substantially test our ability to cooperate on a major project.

As a last point, it is my belief that to meet the no-regrets obligation resources must be available to accomplish something big. Therefore, we need ample supplies of concentrated energy, rare earth materials, advanced technologies (space, etc.), and we need a shared human endeavor that goes beyond myopic international relations. In this way, the global grid could also be seen as a source of insurance that ample energy supplies will be available to meet this perpetual obligation.

Perpetual Obligation #12. Fully inform current generations about futures and obligations

The need to fully inform current generations on the status and prospects of meeting obligations to future generations cannot be overstated. This obligation needs to be shared the world over and communicated as clearly as possible. Meeting perpetual obligations ought not be a subject of partisan politics and campaigns of misinformation. Those responsible for communicating information need to be mindful of the heuristics and biases individuals employ when trying to understand risk and uncertainty.[44] They also need to be aware of the difficulties individuals have with respect to imagining futures even a few years out[45] and how cultural beliefs can constrain futures-thinking.[46] Efforts should be also considered to make futures-thinking approachable and digestible so that individuals can participate in global discussions about future generations.[47] The Global Anticipation and Decision Support System introduced in Chapter 7 can greatly assist with these activities.

Humanity needs to practice effective foresight and anticipation research in order to provide sound insights into how well we may or may not meet our obligations to future generations in the years to come. The point of this entire exercise is to avoid crisis situations that may require a war footing to deal with imminent existential threats when lesser actions implemented decades or even centuries earlier would have been sufficient to deal with the threats.

Metrics for this obligation are relatively straightforward. Approaches used to measure proficiency in math and reading can be adopted and adapted to measure individuals' future literacy. Miller et al. provide an excellent start on how to conceive and measure futures literacy.[48] The SCES and other scenarios introduced earlier could play a very important role in meeting this obligation.

In summary, the set of twelve perpetual obligations is designed to be comprehensive though reasonable minds could certainly reformulate the list and add additional obligations to the set. The twelve obligations should be considered incommensurable. This means that they cannot be compared to each other to determine which is more important. Fundamentally, the obligations are equally important, all are to be invested in, and they are not to be traded off against each other.

On the other hand, programs to meet some perpetual obligations could also help meet others. For example, one could argue that sustainable systems of production and sustainable societies would by themselves reduce risks of human extinction, at least with respect to existential risks categories I–V that are anthropogenically driven. Generation of new knowledge can contribute to the development of more sustainable systems of production. Improving futures literacy of the world's population should assist with meeting all of the other perpetual obligations.

Lastly, it needs to be stressed that meeting perpetual obligations is not an optimization exercise per se. Certainly, it is advisable to meet perpetual obligations in cost-effective manners, but I do not conceive that there are optimal approaches to preventing human extinction nor should we fret over finding perfect approaches. I think that Herbert Simon's concept of satisficing is most appropriate in this context.[49] Perfect solutions are not possible because optimal solutions may not only be unattainable but also indefinable. Satisfactory solutions will get the job done.

Table 6.2 maps previously published statements of obligations of future generations to the twelve perpetual obligations presented earlier. Almost all twelve have been previously mentioned, some more explicitly than others. For example, the maintaining options, no regrets, and sustainability obligations previously addressed are explicitly addressed in the set of twelve perpetual obligations. UNESCO's comprehensive list of obligations maps to the twelve in several manners, from respect to preventing human extinction and extinction of earth life to bequeathing sustainable societies. Overall, the set of twelve introduces one brand new concept, R&D to protect the journey and future generations and makes more explicit the obligations with respect to futures literacy and protecting our knowledge base and the essences of nature and the human species. The set of twelve is very explicit with respect to preventing human extinction and the extinction of earth life. The set of twelve is also more policy-oriented than previous statements.

Table 6.2 Mapping Statements on Obligations to Future Generations to Perpetual Obligations

Source	*Obligation*	*Mapping to Perpetual Obligations*											
		1	2	3	4	5	6	7	8	9	10	11	12
Bell	Regarding the natural resources of the earth, present generations have no right to use to the point of depletion or to poison what they did not create				dark	light							
	Past generations left many of the public goods that they created not only to the present generation but to future generations as well										dark		
	There is a *prima facie* obligation of present generations to ensure that important business is not left unfinished											dark	
Slaughter	The human project is unfinished											light	
	We are partly responsible for the dangers to their well-being					light							
	The global commons has been compromised by human activity and restorative actions are necessary				light	light	light						
McClean; Schrader-Frechette	Fairness obligation with respect to risk					dark							
Weiss; Golding; Tonn; Gilroy	Maintaining options obligation								dark				
Tough	Peace and security			light									
	Environment				light	light							
	Children and learning			light									light
UNESCO #1	Needs and interests of future generations			light									
UNESCO #2	Freedom of choice								dark				
UNESCO #3	Maintenance and perpetuation of humankind	dark			light								
UNESCO #4	Preservation of life on earth		light										
UNESCO #5	Protection of the environment		light			light							
UNESCO #6	Human genome and biodiversity						light	light					
UNESCO #7	Cultural heritage and cultural heritage			light									
UNESCO #8	Common heritage of humankind			light									
UNESCO #9	Peace			light									
UNESCO #10	Development and education			light									light
UNESCO #11	Non-discrimination			light									

Legend: Dark gray – strong linkage; Light gray – moderate linkage; Blank – no linkage.

At a first glance, the list of twelve perpetual obligations does not appear to overlap much if at all with the lists of obligations to future generations proposed by Bell and Slaughter that are provided in Chapter 5. Most of the entries in their lists pertain to why we should care about future generations. However, in a few instances, several assertions in their lists have been translated into perpetual obligations. For example, both writers assert that the current generations have no right to leave a degraded environment to future generations. This assertion is expressed as several perpetual obligations, from bequeathing sustainable systems of production (#4) to reducing extinction risks (#2) to reducing environmental risks (#5). Both authors assert that humanity has not yet achieved all it might. This assertion has been translated into the minimize regrets obligation (#11). Lastly, Bell asserts it is important that public goods enjoyed today ought to also be enjoyed by future generations. This assertion is captured, in part, by obligation to preserve knowledge (#9).

Many of the responsibilities set out by the UNESCO declaration correspond to the list of perpetual obligations. For example, Article 3, maintenance and perpetuation of humankind is another way of stating the prevention of human extinction obligation (#1). Article 4 is a different way to state the prevention of the extinction of earth life (#2). Article 2, freedom of choice, can be tied to the maintaining options obligation (#8). I believe, though, that how the twelve perpetual obligations are stated makes them more amenable to measurement and policy assessment than those proposed by UNESCO.

Reasons why we should care about future generations and what our obligations are to future generations should be considered timeless and universal, in other words perpetual. While well-meaning individuals may disagree on the specifics of why and what, the framework woven earlier assumes that all generations would agree that why and what are important. These values are immutable, not relative. Belief systems that may hold that why and what are not important at all are not defensible.

Summary

Table 6.3 presents my personal, subjective judgment as to how well current generations are meeting each of the dozen perpetual obligations. As captured in the table, I am extremely concerned about the lack of sustainability in current modes of production, and that individuals around the world are exposed to environmental risks that are much too high. I am also worried about protecting the essence of nature and of humans because technology is evolving so fast and there are virtually no serious efforts outside of the writings of a relatively few academics that are addressing these issues. My cursory review of R&D expenditures worldwide suggests that the perpetual obligation #9 is not being met. Humanity is not engaged in a shared, no-regrets, unfinished business project, like building WEMS or a global grid. Futures literacy across the globe is almost non-existent. The emergence of

Table 6.3 Meeting Our Perpetual Obligations: Scorecard

Perpetual Obligation	*Assessment*	*Comments*
1. Prevent Human Extinction	F	Several plausible SCES have been developed and extinction-event categories VI–IX are not being seriously addressed. The probability of human extinction likely exceeds 10^{-20}.
2. Prevent Extinction of Earth Life	C	Serious risks threaten earth life but none seems capable of completely extinguishing all life on earth in the next 1000 years.
3. Bequeath Sustainable Societies	C	Mixed. Progress has been made toward achieving sustainable societies, but these achievements are precarious.
4. Bequeath Sustainable Production	F	Despite the embrace of sustainability worldwide, production systems are still drawing down resources and emitting pollutants faster than the earth can deal with.
5. Prevent Environmental Risks	F	Too many individuals in too many regions suffer from unacceptable environmental risks, from air and water pollution, to increasing risks from climate change.
6. Protect Essence of Nature	F	Humans have irrevocably altered the earth's ecosystems and we are in the midst of a sixth mass extinction. We have not adequately thought through the use of genetic engineering to 'protect' the essence of nature.
7. Protect Essence of Humans	F	We seem to be quickly moving toward a transhuman society without serious discussion of the ethical implications.
8. Maintain Options	C	Numerous problems listed earlier will continue to constrain our choices regarding production, but in most regions of the world culture seems to be free to evolve.
9. Conduct Futures R&D	F	Very few R&D resources are devoted to improving our long-term survival as a species.
10. Protect Knowledge	C	The world is in the process of digitization, where even ancient knowledge artifacts are being stored in the cloud. The issue is whether the cloud itself is resilient over the long term.

(*Continued*)

Table 6.3 (Continued)

Perpetual Obligation	*Assessment*	*Comments*
11. Minimize Regrets	F	We are rapidly drawing down resources that we could need to achieve great things. Humanity is not engaged in a shared, no-regrets project.
12. Fully inform current generations	F	We have made almost no progress in this area.

Legend: Blank – Meeting; Light gray – Concerned; Dark gray – Failing.

the 'cloud' provides important seeds to protect human knowledge but until explicit plans are in place to protect human knowledge, this obligation will not be met.

Ethical risk threshold of human extinction

This last section of Chapter 6 addresses in depth Perpetual Obligation #1: preventing human extinction. The first task is to present the approach used to establish the 10^{-20} threshold that has already been alluded to several times earlier. Then, the second task is to assemble several pieces of evidence to use to judge whether humanity is currently meeting or exceeding the threshold.

The biggest picture theme of this book leads to this question: what is the acceptable risk threshold of human extinction? This question is an important one in the field of risk analysis. One of the seminal papers in this field asks a slightly different question: How Safe is Safe Enough?[50] From the perspective of the field of risk analysis, one can ask this question about many things. For example, how safe should a nuclear power plant be? How safe should our roads be? Air travel? Our air? Our water? The crux of the question is that virtually nothing can be made completely risk free and at some point, incrementally improving safety can entail exponentially increasing costs.

The question of how safe is safe enough has never been definitively answered. However, a *de facto* standard of 10^{-6} has emerged over the years as noted in the discussion of Perpetual Obligation #5. This standard means that a specific involuntary environmental risk ought not result in the deaths of more than one in a million individuals in a given year. In other words, the probability of an individual dying who is exposed to this risk, say drinking water tainted with arsenic, ought to be less than one in a million. This means that the environmental risks ought to be studied, individuals' exposure to the risks need to be estimated, mortality dose-response curves need to be estimated, and if the involuntary risks to the population is greater than 10^{-6}, regulations need to be adopted to reduce the risks to below this ethical threshold.

Here is a start to answering the big picture question related to an ethical risk threshold for human extinction. The absolute minimum should be 10^{-6}. However, this threshold is for one individual during that person's lifetime. One can argue that the threshold for human extinction should be much more stringent. The threshold I developed is 10^{-20} and here, briefly is how I came to this estimate.[51]

First, I needed to set some large-scale parameters for the exercise. I decided that the steady-state human population in question would be ten billion individuals and that life on earth would last another one billion years, consistent with our serendipitous trip through the future in Chapter 2.

Second, I needed to adopt some guiding principles, in this case, three of the twelve perpetual obligations that appeared most amenable to this exercise: Perpetual Obligations #5, 8, and 11, relating to involuntary environmental risks, maintaining options, and no-regrets/unfinished business, respectively. Each of these perpetual obligations is operationalized to produce an ethical threshold of human extinction.

Let's start with Perpetual Obligation #5. As discussed earlier, the fairness principle holds that no one in the future ought to be exposed to higher involuntary environmental risks than is deemed ethically acceptable for current generations. And, as also mentioned the *de facto* standard is 10^{-6}. In our case, we are concerned about risks to all ten billion individuals rather than just one. Thus, at a minimum, ethical threshold for human extinction should be ten billion times more stringent, 10^{-16}.

The calculation is more complicated, however, because the risk threshold of 10^{-16} is actually the risk threshold for the very last generation of humans on the earth, just short of one billion years from now. To get that far into the future, the risk threshold for time periods leading up to this last time period needs to be more stringent.[52] I calculated that for current generations, the ethical threshold is actually much higher, around 10^{-24}. Over time, the risk threshold tied to this obligation declines to 10^{-16}.

Perpetual Obligation #11 embodies the no-regrets notion proposed by Wendell Bell that humans will always have unfinished business and it would be a shame if we went extinct before we accomplished all there is to accomplish. My interpretation of the unfinished business obligation is that our species has many very significant goals to accomplish which may take many, many more years, maybe even millions of years. I also believe that many, most or even all of the things that we need to accomplish as a species have not yet been revealed to us. Thus, we should regret becoming extinct not only because we have not accomplished all we could, but we also need to be around to better understand what those goals could be.

There is a concept in the field of decision analysis called regret.[53] Depending on the situation being faced, regret could weigh heavily upon a decision. For example, someone may choose a more difficult path (e.g., career, sports, food) so as not to feel regret years later over not choosing the path. In this way, regret can be tied to decision situations that have substantial amounts of uncertainty. The benefits of the chosen path are uncertain but intriguing, certainly more intriguing than taking a well-worn path.

The challenge is how to develop an ethical risk threshold of human extinction from this perpetual obligation. The first important assumption is that regret increases over time as one sees less time over the horizon to finish unfinished business. Anxiety increases as the clock is running out. It is assumed, as noted earlier, that we have one billion years to finish our unfinished business. For this exercise, it is assumed that we have not yet finished this business in one billion years, which means that regret increases over time as we near the end of life on earth. I developed a mathematical function that expresses increasing regret over time.[54] When these assumptions and mathematical assumptions are combined, the resulting ethical threshold of human extinction for the current generations is approximately 10^{-20}. This risk threshold increases over time.

Perpetual Obligation #8 holds that current generations ought not confine future generations' freedom to shape their lives and cultures as they may.[55] My primary interpretation of this obligation is that current generations ought not bequeath to future generations extraordinarily unsustainable systems of production and diminished global ecosystems such that future generations can only follow a path or two into the future simply to survive.

For the purposes of establishing an ethical threshold for the risk of human extinction, the issue here is that current generations will need to invest resources to ensure not only the ability of future generations to live but also to have freedom to shape their lives. Sometimes, current generations will need to invest little and sometimes a lot, depending upon current and anticipated existential risks. Another mathematical function was developed to capture how much should be invested depending on the level of existential risk. At a reasonable 5% level of societal investment to maintain options, the function indicates that this level of investment is appropriate if the probability of human extinction is 10^{-20}. If this level of investment is assumed to be constant across time, then this particular threshold also stays constant over time.

In summary, three very different approaches were taken to establish an ethical threshold for the risk of human extinction:

- Fairness obligation – 10^{-24}, decreases over time
- No regrets obligation – 10^{-20}, increases over time
- Maintain options obligation – 10^{-20}, constant over time

For purposes of discussion, let's assume that the risk level triangulates to 10^{-20}. Thus, we now have a policy-relevant guidepost with respect to human extinction generations in the early parts of the third millennium. In other words, policies need to be designed, implemented, administered, and, of course fairly evaluated to ensure that the risk of human extinction does not exceed this threshold. Action must be taken if the threshold is currently exceeded or anticipated to be exceeded. These actions are addressed in Chapter 7.

Is this threshold currently being exceeded? It should also be stated that no one really knows the current probability of human extinction, and there is no one universally accepted manner to estimate the said probability.[56] However, six signals are drawn upon to support the verdict that the answer probably is yes. The first signal is found in Chapter 3, where it is argued that extinction has been the fate of the vast majority of species on the earth and maybe also most if not all intelligent life in the universe.

The second signal is that many authors cited in that chapter argue the probability of human extinction is quite likely in the short term, especially with respect to the 10^{-20} threshold.

The third signal is captured in Table 6.4. Contained in this table are estimated probabilities for a selection of extinction-level events, also introduced in Chapter 3. Certainly, it is challenging to develop most of these estimates, given the paucity of empirical data we have to work with, say, on tears in space–time continuums. Thus, it cannot be argued that these estimates are extremely precise. However, it can be argued that all of the estimates are quite larger than the 10^{-20} ethical threshold of extinction risk just developed.

As discussed in Chapter 4, the probabilities of extinction-level events are not actual probabilities of human extinction for a couple of reasons. Thus, the fourth signal is that humans are resilient. Even without the benefit of anticipation and planning to avoid our own extinction, it is very likely that humanity will survive pandemics and even nuclear war, though the risk of catastrophic loss of life cannot be ignored and such risk may be quite higher than ethically acceptable. History also suggests that failing these provisions,

Table 6.4 Estimated Probabilities of Some Extinction-Level Events

Extinction-Causing Event	*Annual Probability of Event Occurring*
Non-friendly Super-AI[57]	5×10^{-2}
Carrington Class Ejection from the Sun[58]	5×10^{-2} to 2×10^{-1}
Pandemic[59]	4×10^{-2}
Nuclear War[60]	1×10^{-2}
Runaway Nanotechnology[61]	5×10^{-3}
Super-Volcano Eruption[62]	2×10^{-5}
Gamma Ray Burst in Milky Way[63]	5×10^{-5}
Extinction-Level Asteroid Impact[64]	1×10^{-7}
Tear in Space–Time Continuum from High-Energy Physics Experiment[65]	2×10^{-8}
Cosmic Sterilization Rate (includes Near Earth Super or Hypernova; Rogue Black Hole; Vacuum Phase Transition)[66]	1×10^{-9}

that some humans, if even only a few thousand in number, will probably survive even a super-volcanic eruption.[67]

The fifth signal is we are capable of anticipating threats to our own existence and implementing plans and programs. For example, humans have developed procedures for containing contagious viral and bacterial infections and rudimentary capabilities for quickly manufacturing treatments and vaccines.[68] To deal with catastrophic climate change, humans could live in completely self-sufficient habitats scattered across the globe, or even in nearby space, as suggested by Stephen Hawking.[69] Unfortunately, humanity is not actively anticipating extinction-level events nor implementing plans or programs to deal with almost all of the events. We are also lacking in our approaches to dealing with CEs, UUCs, and UKUKs.

Lastly, we have extinction scenarios presented at the end of Chapter 3 and the SCES presented at the end of Chapter 4 to work into our assessment. The SCES presented at the end of Chapter 4 was estimated to have about a 10^{-20} or lower chance of occurring. The autonomous vehicle and gene drive scenarios appear to be at least as unlikely. However, it only takes two extinction scenarios with a probability of 10–20 to exceed the ethical threshold. Are there two or more extinction paths for every 10^{20} paths moving 1000 years into the future or one or less (say one every 10^{24} paths)?

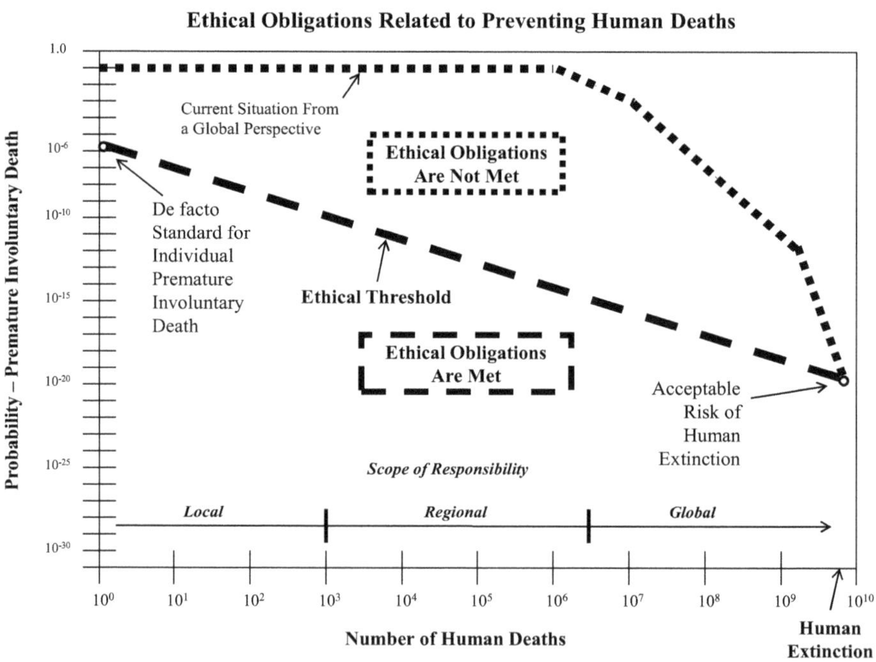

Figure 6.1 How Well is Humanity Meeting Its Ethical Obligations to Prevent Human Extinction[70]

At this point, the evidence is inconclusive though, given these six signals, it seems reasonable to argue that we are close to exceeding the ethical threshold. Figure 6.1 illustrates this. The dashed line represents the ethical threshold. On the left-hand side, it starts at the probability of 10^{-6} for one death from an involuntary risk and then extends downward till the 10^{-20} probability of human extinction is reached. Probabilities above the dashed line exceed ethical thresholds whereas probabilities below the dashed line are ethically satisfactory. Figure 6.1 suggests that the probability of human extinction, captured at the end of the dotted line, may be just at the ethical threshold. On the other hand, the figure suggests that risks of individual deaths and global catastrophic loss of life are almost certainly above the dashed line.

Notes

1 E. Partridge (Ed.). 1981. *Responsibilities to Future Generations: Environmental Ethics.* Prometheus Books, Buffalo, NY.
2 W. Bell. 1993. Why Should We Care About Future Generations? In H. Didsbury (Ed.), *The Years Ahead: Perils, Problems, and Promises.* World Future Society, Washington, DC, 25–41; R. A. Slaughter. 1994. Why We Should Care for Future Generations Now. *Futures,* 26, 1077–1085.
3 D. MacLean, D. Bodde, and T. Cochran. 1981. *Introduction to Conflicting Views on a Neutrality Criterion for Radioactive Waste Management.* Center for Philosophy and Public Policy, University of Maryland, College Park, MD.
4 B. E. Tonn. 1987. Philosophical Aspects of 500-year Planning. *Environment and Planning A,* 20, 1507–1522.
5 K. Schrader-Frechette. 1991. Ethical Dimensions and Radioactive Waste. *Environmental Ethics,* 13, 327–344.
6 E. B. Weiss. 1989. *In Fairness to Future Generations: International Law, Common Patrimony, and Intergenerational Equity.* Transnational Publishers Inc., Dobbs Ferry, New York.
7 B. E. Tonn. 1987, ibid.
8 M. P. Golding. 1981. Obligations to Future Generations. In E. Partridge (Ed.), *Responsibilities to Future Generations: Environmental Ethics.* Prometheus Books, Buffalo, NY, 61–72.
9 J. Balazs and J. Gaspar. 2010. Taking Care of Each Other: Solid Economic Base for Living Together. *Futures,* 42, 69–74.
10 B. E. Tonn. 1986. 500-year Planning: A Speculative Provocation. *Journal of the American Planning Association,* 52, 2, 185–193.
11 M. P. Golding. 1981, ibid.
12 J. Gilroy. 1992. Public Policy and Environmental Risk: Political Theory, Human Agency, and The Imprisoned Rider. *Environmental Ethics,* 14, 3, 217–237.
13 A. Tough. 1993. What Future Generations Need from Us. *Futures,* 25, 10, 1041–1050.
14 http://portal.unesco.org/en/ev.php-URL_ID=13178&URL_DO=DO_TOPIC&URL_SECTION=201.html
15 Text that describes the twelve perpetual obligations is adapted from B. E. Tonn. 2017. Philosophical, Institutional, and Decision Making Framework for Meeting Obligations to Future Generations. *Futures,* 95, 44–57.
16 T. Ord. 2019. In J. Brockman (Ed.), *The Last Unknowns.* William Morrow, New York, 212.

17 See the discussion about the Global Anticipation and Decision Support System in Chapter 7.
18 B. E. Tonn. 2007. Futures Sustainability. *Futures*, 39, 1097–1116.
19 E. O. Wilson. 1984. *Biophilia – The Human Bond with Other Species*. Harvard University Press, Cambridge, MA.
20 B. McKibbon. 2003. *Enough: Staying Human in an Engineered Age*. Henry Holt & Co., New York.
21 B. E. Tonn. 1999. Transcending Oblivion. *Futures*, 31, 351–359.
22 B. E. Tonn. 2009. Preventing the Next Mass Extinction: Ethical Obligations. *Journal of Cosmology*, 2, 334–343.
23 E. Victor and L. Guidry-Grimes. 2014. The Persistence of Agency Through Social Institutions and Caring for Future Generations. *International Journal of Feminist Approaches to Bioethics*, 7, 1, 122–141.
24 J. Thompson. 2017. The Ethics of Intergenerational Relationships. *Canadian Journal of Philosophy*, 47, 2–3, 313–326.
25 B. Barber. 1994. *Strong Democracy: Participatory Politics for a New Age*. University of California Press, Berkeley, CA.
26 https://freedomhouse.org/report/freedom-world/freedom-world-2019
27 www.cato.org/human-freedom-index-new
28 https://en.wikipedia.org/wiki/World_Index_of_Moral_Freedom
29 www.hh.se/english/research/research-environments/research-on-education-and-learning-clks/research-projects-within-clks/maxrange—analysing-political-regimes-and-democratisation-processes.html
30 www.eiu.com/topic/democracy-index?&zid=democracyindex2019&utm_source=blog&utm_medium=blog&utm_name=democracyindex2019&utm_term=democracyindex2019&utm_content=middle_link
31 D. Meadows. 1994. Envisioning a Sustainable World. Third Biennial Meeting of the International Society for Ecological Economics, San Jose, Costa Rica, October 24–28.
32 W. Steffen, K. Richardson, J. Rockström, S. Cornell, I. Fetzer, E. Bennett, et al. 2015. Planetary Boundaries: Guiding Human Development on a Changing Planet. *Science*, 347, 6223.
33 www.climate.gov/maps-data/primer/climate-models; www.iea.org/reports/world-energy-model; www.epa.gov/climate-research/modeling-interactive-effects-nitrogen-deposition-and-climate-change-terrestrial
34 P. Hunter and L. Fewtrell. 2001. Acceptable Risk. In L. Fewtrell and J. Bartram (Eds.), *Water Quality: Guidelines, Standards and Health*. World Health Organization, IWA Publishing, London; K. Kelly. 1991. *The Myth of 10^{-6} as a Definitive Acceptable Risk*. Presented at 84th Annual Meeting Air & Waste Management Association, Vancouver, BC, Canada, June 16–21.
35 P. Vitousek, H. Mooney, J. Lubchenco, and J. M. Melillo. 1997. Human Domination of Earth's Ecosystems. *Science*, 277, 494–499.
36 https://en.wikipedia.org/wiki/CRISPR
37 https://ensia.com/features/crispr-biodiversity-coral-food-agriculture-invasive-species/
38 J. Diamond. 1997. *Guns, Germs and Steel*. W.W. Norton, New York.
39 N. Bostrom. 2005. Transhumanist Values. *Review of Contemporary Philosophy*, 4, 87–101.
40 R. Kurzweil and T. Grossman. 2005. *Fantastic Voyage: Live Long Enough to Live Forever*. Rodale, New York.
41 B. E. Tonn. 2004. Research Society: Science and Technology for the Ages. *Futures*, 36, 335–346.
42 W. Bell. 1993, ibid.

43 www.geni.org/globalenergy/issues/overview/grid.shtml
44 A. Tversky and D. Kahneman. 1973. Availability: A Heuristic for Judging Frequency and Probability. *Cognitive Psychology*, 5, 207–232; A. Tversky and D. Kahneman. 1974. Heuristics and Biases: Judgement Under Uncertainty. *Science*, 185, 1124–1130; D. Kahneman and A. Tversky. 1979. Prospect Theory: An Analysis of Decision under Risk. *Econometrica*, 47, 2, 263–292; P. Slovic. 1987. Perception of Risk. *Science*, 236, 280–285.
45 B. E. Tonn, F. Conrad, and A. Hemrick. 2006. Cognitive Representations of the Future: Survey Results. *Futures*, 38, 810–829.
46 B. E. Tonn and D. MacGregor. 2009. Individual Approaches to Futures Thinking and Decision Making. *Futures*, 41, 117–125; B. E. Tonn. 2005. Thinking About the Future: Observations from the Field. *Futures Research Quarterly*, 20, 4, 33–46.
47 B. E. Tonn and D. Stiefel. 2012. The Future of Governance and the Use of Advanced Information Technologies. *Futures*, 44, 812–822.
48 Riel Miller (Ed.). 2018. *Transforming the Future: Anticipation in the 21st Century*. Routledge, New York.
49 H. Simon. 1979. Rational Decision Making in Business Organizations. *American Economic Review*, 69, 4, 493–513.
50 B. Fischhoff, P. Slovic, S. Lichtenstein, S. Read, and B. Combs. 1978. How Safe Is Safe Enough? Psychometric Study of Attitudes Towards Technological Risks and Benefits. *Policy Sciences*, 9, 127–152.
51 For an extended discussion, see B. E. Tonn. 2009. Obligations to Future Generations and Acceptable Risks of Human Extinction. *Futures*, 41, 427–435.
52 Here is an example to help explain this. Let's assume you have an experiment to do and you need to convince your supervisor that there is at least a 50% chance that the experiment will be a success. The experiment consists of three sequential steps and each step needs to be successfully implemented. If each step has a 50% chance of success, then the probability that the experiment will be successful is .5×.5×.5= .125. The probability of success is way too low at this point, so the probabilities of each step need to be higher than the end goal probability. A good goal would be to strive to ensure that each step has an 80% chance of success, which yields a .51 probability that the experiment will be successful (.8×.8×.8=.51). The same process is at work with respect to this estimate of an ethical threshold of human extinction.
53 G. Loomes and R. Sugden. 1982. Regret Theory: An Alternative Theory of Rational Choice Under Uncertainty. *The Economic Journal*, 92, 368, 805–824; D. Bell. 1982. Regret in Decision Making Under Uncertainty. *Operations Research*, 30, 5, 961–981.
54 See B. E. Tonn. 2009, ibid., Figure 3.
55 E. Partridge (Ed.). 1981. *Responsibilities to Future Generations: Environmental Ethics*. Prometheus Books. Buffalo, NY.
56 B. E. Tonn and D. Stiefel. 2013. Evaluating Methods for Estimating Existential Risks. *Risk Analysis*, 33, 10, 1772–1787.
57 www.webcitation.org/6YxiCAV0p?url=www.fhi.ox.ac.uk/gcr-report.pdf
58 www.nature.com/articles/s41598-019-38918-8#citeas
59 T. Day, J-B. André, and A. Park. 2006. The Evolutionary Emergence of Pandemic Influenza. *Proceedings of the Royal Society | Biological Sciences*, 273, 1604, 2945–2953.
60 www.webcitation.org/6YxiCAV0p?url=www.fhi.ox.ac.uk/gcr-report.pdf
61 Ibid.
62 B. Harris. 2008. The Potential Impact of Super-volcanic Eruptions on the Earth's Atmosphere. *Royal Meteorological Society*, 63, 8, 222–225. https://doi.org/10.1002/wea.263.

63 A. L. Melott, B. S. Lieberman, C. M. Laird, L. D. Martin, M. V. Medvedev, B. C. Thomas, J. K. Cannizzo, N. Gehrels, and C. H. Jackman. 2004. Did a Gamma-ray Burst Initiate the Late Ordovician Mass Extinction? *International Journal of Astrobiology*, 3, 55–61. arXiv:astro-ph/0309415. Bibcode 2004IJAsB . . . 3. . . 55M. https://doi.org/10.1017/S1473550404001910.

64 C. Chapman. 2004. The Hazard of Near-earth Asteroid Impacts on Earth. *Earth and Planetary Science Letters*, 222, 1–15.

65 W. Busza, R. Faffe, J. Sandweiss, and. F. Wilczek. 1999. *Review of Speculative "Disaster Scenarios" at RHIC.* Brookhaven National Laboratory Report, September, Upton, NY, 28. Retrieved February 26, 2012, from www.arxiv.org/abs/hep-ph/9910333v1. A. Dar, A. De Rujula, and U. Heinz. 1999. Will Relativistic Heavy Ion Colliders Destroy Our Planet? *Physical Letters*, B470, 142–148.

66 M. Tegmark and N. Bostrom. 2005. Is a Doomsday Catastrophe Likely? *Nature*, 438, December 8, 754–756.

67 https://en.wikipedia.org/wiki/Toba_catastrophe_theory

68 Please see the Epilogue for specific comments about COVID-19.

69 www.telegraph.co.uk/science/2017/06/20/human-race-doomed-do-not-colonise-moon-mars-says-stephen-hawking/

70 This figure is adapted from B. E. Tonn. 2009, ibid., 334–343.

7 Frameworks and tools for actions-oriented futures policy making

The preceding chapters establish that humanity faces numerous risks, many of which are by themselves extinction-level risks and many of which can link into chains of events that could also cause the demise of humanity. We ought to care about reducing these risks to acceptable levels. We can do so, in part, by working diligently to meet our perpetual obligations. The next logical topic to tackle is what actions humanity should take to address unmet perpetual obligations.

The first section of this chapter presents a table that lists metrics for determining to what extent each of the perpetual obligations are being met. Then, the second section maps these gradations to a big picture policy action framework. This framework has six action categories that humanity can take to address meeting the twelve perpetual obligations. How to deal with uncertainties related to mapping obligations to the framework is addressed in the third section using human extinction as an example. My subjective judgments as to where each perpetual obligation lays with respect to each action level is presented and discussed in the fourth section.

The fifth section introduces the concept of 1000-year planning to help guide humanity's long-term policies and programs to meet our perpetual obligations. The rest of the chapter focuses on sub-perpetual obligations-level global goals and sub-global goals, policies, and programs whose impacts in the aggregate will determine whether humanity is meeting its perpetual obligations. The United Nation's Sustainable Development Goals (SDGs) represent a set of goals of this nature. National transportation plans and local land use ordinances are examples of sub-global policies and programs that should be synchronized with 1000-year planning objectives and with each other. To help with this coordination effort, a second new system is proposed, simply called the Global Anticipation and Decision Support System (GADSS). This system would be designed to help decision makers at all levels of government and those in the non-profit and private sectors anticipate challenges to meeting perpetual obligations and also anticipate everybody else's decisions with respect to these challenges and everyone else's current generation-focused decisions.

Metrics to guide justifiable actions to meet perpetual obligations[1]

Table 7.1 presents suggestions that set metric-related quantitative and qualitative thresholds that link each perpetual obligation to each action category. Each row of the table expresses the extent to which current generations are meeting each particular perpetual obligation. Referring to the discussion in Chapter 6 about the ethical risk threshold of human extinction, the first cell in Perpetual Obligation #1's row indicates that this obligation is being met if both the upper and lower probabilities of human extinction fall below 10^{-20}. The next five cells indicate situations where the extent of not meeting this obligation increases from very small to small, moderate, large, and extremely large.

Table 7.1 contains several nuances that need to be highlighted and explained. First, the unshaded cells in the table indicate preferred states. Thus, as already explained, the first cell in the row for Perpetual Obligation #1 is the metric for the preferred state of that perpetual obligation. Similarly, this is the same case for Perpetual Obligations #2–#8. This is not the case for Perpetual Obligations #9–#12, where the second and third cells represent the preferred states. This nuance captures the judgment that humanity need not devote resources to dealing with the first eight perpetual obligations if they are being met but that some resources should always be devoted to the other four perpetual obligations. This nuance is addressed in more depth in the next section.

A second nuance is that as the extent to which we are not meeting Perpetual Obligations #3–#7 increases, at some point humanity will be faced with high levels of risks associated with human extinction and the extinction of life on earth, Perpetual Obligations #1 and #2, respectively. For example, it is certainly plausible for the probability of involuntary death of individuals from environmental risks could increase to where the cumulative risks threaten human extinction. Thus, it makes sense from a policy perspective to address combinations of perpetual obligations as the seriousness of not meeting them increases. Several other nuances are discussed in the next section.

Policy – big picture action framework

Actions need to be balanced against risks. The higher the risks, defined generally as the probability of an event times the magnitude of the consequence, the more that actions need to ramp up exponentially in scale and scope. This section addresses the top row of Table 7.1. Here is found a six-component framework to help guide humanity's decisions about what to do if we are not meeting our perpetual obligations. The framework for societal actions has these six levels: I. Do Nothing; II. Policy Nudges; III. Major Programs; IV. Manhattan Project-Scale Programs; V. Ubiquitous Command and Control Regulations; and VI. War Footing. Figure 7.1 illustrates the

Table 7.1 Linking Perpetual Obligations to the Framework of Global Actions

Obligations/ Justifiable Actions	*I. Do Nothing*	*II. Policy Nudges*	*III. Major Programs*	*IV. Manhattan-Scale Projects*	*V. Ubiquitous Command & Control*	*VI. War Footing*
1. Prevent human extinction	$UP << 10^{-20}$ $LP << 10^{-20}$	$UP > 10^{-20}$ $LP < 10^{-20}$	$UP > 10^{-20}$ $LP > 10^{-20}$	$UP >> 10^{-20}$ $LP > 10^{-20}$	$UP >> 10^{-20}$ $LP >> 10^{-20}$	$UP >>> 10^{-20}$ $LP >>> 10^{-20}$
2. Prevent extinction of life on earth	$UP << 10^{-23}$ $LP << 10^{-23}$	$UP > 10^{-23}$ $LP < 10^{-23}$	$UP > 10^{-23}$ $LP > 10^{-23}$	$UP >> 10^{-23}$ $LP > 10^{-23}$	$UP >> 10^{-23}$ $LP >> 10^{-23}$	$UP >>> 10^{-23}$ $LP >>> 10^{-23}$
3. Bequeath sustainable societies	High quality of life, rights protected, stable and wise governance	Significant social justice issues	Major social justice issues, civil unrest	Many regions unstable, widespread injustice	Human civilization is at risk	Merges into human extinction risk
4. Bequeath sustainable production systems	$UP = 1.0$ $LP = 1.0$	$UP < .5$ $LP < .1$	$UP < .1$ $LP < .01$	$UP < .01$ $LP < .001$	$UP < .001$ $LP < .0001$	Merges into human extinction risk
5. Reduce death from involuntary environmental risks	$UP << 10^{-6}$ $LP << 10^{-6}$	$UP > 10^{-6}$ $LP < 10^{-6}$	$UP >> 10^{-6}$ $LP >> 10^{-6}$	$UP >>> 10^{-6}$ $LP >>> 10^{-6}$	Merges into human extinction risk	Merges into human extinction risk
6. Preserve essence of nature	All nature essentially un-engineered	Numerous ethical situations exist	Augmented species dominant in nature	Augmentation causing serious unintended consequences	Merges into species extinction risk	Merges into species extinction risk

(*Continued*)

Table 7.1 (Continued)

Obligations/ Justifiable Actions	*I. Do Nothing*	*II. Policy Nudges*	*III. Major Programs*	*IV. Manhattan-Scale Projects*	*V. Ubiquitous Command & Control*	*VI. War Footing*
7. Preserve essence of human species	All humans essentially un-augmented	Numerous ethical situations exist	Transhumans, robots, AIs widespread	Natural humans close to minority	Merges into human extinction risk	Merges into human extinction risk
8. Maintain options	No regulations or expenditures to meet obligations	Investments needed to spur creativity	Major efforts to address stagnation, dogma	N/A	N/A	N/A
9. Preserve human knowledge	N/A	All data, intellectual work being preserved	Major transition in preservation approach needed	Crisis situation, need large effort to save archives	N/A	N/A
10. Support generation of new knowledge	N/A	0.5% of workforce; 5% of GDP, well-funded	Impending crisis, 1.0% workforce	Crisis event, 1.5% workforce	N/A	N/A
11. Finish important business	N/A	N/A	Active space programs, art, basic science, other	Need large effort to achieve next level opportunities	Take advantage of once-in-a-million-year opportunity	Take advantage of once-in-a-billion-year opportunity
12. Fully inform current generations	N/A	Fully informed with respect to obligations 1–11; robust foresight	Impending crisis ramps up activities	N/A	N/A	N/A

Legend: UP – upper probability; LP – lower probability; > greater than; >> much greater than; >>> very much greater than; < less than; << much less than; <<< very much less than; Blank cell – preferred state; Light gray cell; N/A – not applicable; Dark gray cell – allowable actions.

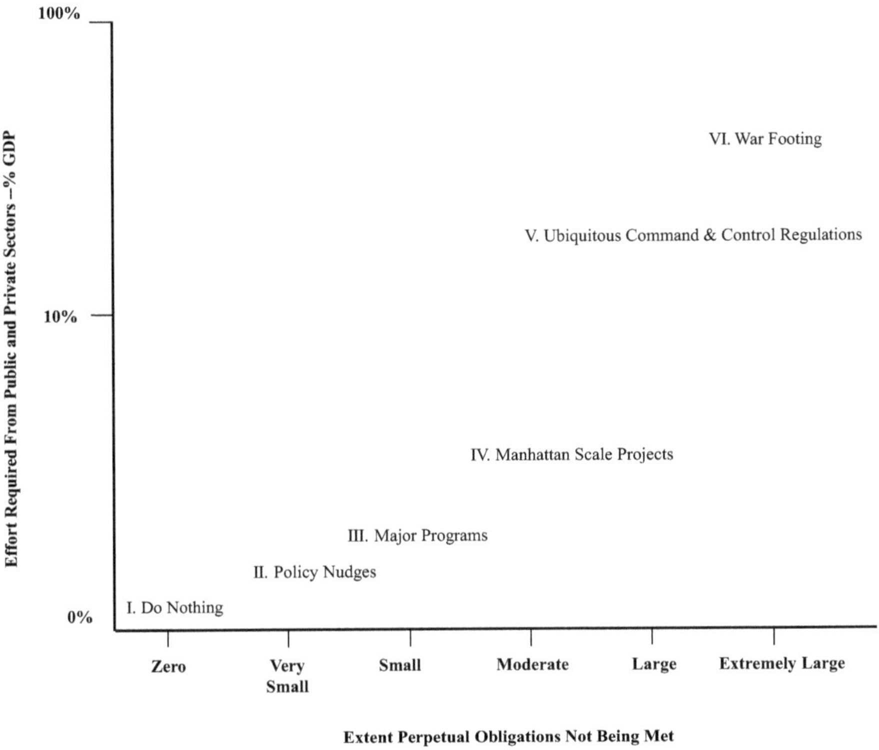

Figure 7.1 Framework for Global Action When Perpetual Obligations Are Not Being Met

relationships of these six levels of action with respect to their potential draw on a country's Gross Domestic Product (GDP) – the aggregate costs for various projects and programs required to accomplish that level of action with the United States in mind as an example – and to the extent to which each perpetual obligation is not being met. Each level is described in more detail in the following pages.

Level I. Do nothing (estimated 0% of GDP)

The Do Nothing component of the societal action framework is self-explanatory and also requires no societal resources. As indicated in Table 7.1, this category of action is preferred with respect to the first eight perpetual obligations if each is being met. Do Nothing is not the preferred option with respect to Perpetual Obligations #s 9–12. In these cases, humanity always need to be engaged in some activities, be they futures literacy programs, or programs to preserve human knowledge.

Level II. Policy Nudges (estimated > 0% to .5% of GDP)

Policy Nudges are the least intrusive and least costly actions to deal with situations where perpetual obligations are not being met. Policy nudges include tax policy, minor research programs, development and deployment programs, and relatively minor regulatory measures. Taxes on emissions of pollutants and tax incentives to promote renewable energy, affordable housing, and energy efficiency are examples of policy nudges. Deployment programs are relatively small efforts funded by governments to improve the energy efficiency of housing or recycle municipal wastes, for example. Regulatory measures include standards (e.g., vehicle fuel efficiency standards), and plans of various sorts (e.g., plans to improve regional watershed quality). While many of these actions may face political resistance, overall the relatively limited scope and scale of these types of actions are usually politically palatable.

Level III. Major programs (estimated 0.5% to 1% of GDP)

The major programs and major public investments level of societal action raises the commitment to between 0.5% to 1% of GDP. For example, a national carbon tax or cap and trade policy, combined with major investments in clean energy technologies, carbon sequestration technologies, comprehensive building retrofits, and major new investments in public transit systems could, combined, represent such a level of investment.

Level IV. Manhattan Project-scale projects (estimated 1% to 5% of GDP)

Thomas Friedman and others have called on the United States to implement a Manhattan Project-scale effort to jointly address energy independence and climate change issues facing the world.[2] The Manhattan Project was the effort by the United States to build the atomic bomb during World War II. Although the Manhattan Project cost around 1% of GDP in its peak year,[3] one can envision multiple projects in this category across the globe that would have a cumulative cost that is 5% of GDP and would be so large that they would dominate the world's attention. In this context, a Manhattan-scale project could be the colonization of space, advocated by some as one way to ensure that humans will not become extinct.[4] Another controversial large-scale project could be a geoengineering solution to avert or reverse climate change.[5] This level of societal action would require the participation of almost every sector of society and the economy of every country on the earth.

Level V. Ubiquitous command and control (estimated 5% to 30% of GDP)

The fifth level requires the imposition of major controls upon society and the economy designed to reduce emissions, energy consumption, and any other human behaviors that are directly contributing to unacceptable levels of human extinction risk. China's one-child policy could be seen as an example of this level of action, as could strict rationing and control of systems of production.

Level VI. War footing (estimated 30% to 50% of GDP)

If humanity is facing a clear and present massive threat to its existence, then one could argue that societies ought to completely mobilize to deal with the threat via an extreme war footing and an economy organized around reducing human extinction risk. This is analogous to a country completely mobilizing in times of war to fight for its own existence. Complete mobilization entails substantial control and direction over the manufacture of technologies needed to deal with existential threats. As a benchmark, in 1943, the United States consumed 48% of its GDP to fight World War II.[6] Human society could be faced with decades and even centuries of extreme war footing depending on the nature of the human extinction risks.

Assessment of risk based on data from a WEMS and integrated SCES would then allow election of appropriate action by the organizations discussed in the next chapter and using the Global Anticipatory Decision Support System presented at the end of this chapter. This table is an important part of the overall blueprint being developed in this book. Much work is needed to polish the contents of the cells and then much more work is needed to develop the methods to estimate upper and lower probabilities directly and through evidential reasoning.

Let's use the third row of Table 7.1 as an example. This row contains cells that link meeting perpetual obligation #3, Bequeath Sustainable Societies, to action categories I through V. For instance, policy nudges would be warranted if the evidence suggests that significant social justice issues exist around the world whereas Manhattan-scale efforts might be needed if many regions around the world are politically and socially unstable and injustice is widespread.

It should be noted that probabilities are the metrics of choice for Perpetual Obligation #4 – Bequeath Sustainable Production Systems. This metric, then, is focused on judging the probability that current and anticipated production systems can be sustained over the 1000-year planning horizon. The Do Nothing action level is justifiable if there is high confidence that yes, production systems are indeed sustainable. Very low levels of confidence that systems are sustainable suggest that substantial actions are needed to meet this perpetual obligation.

A few final thoughts about Table 7.1 and Figure 7.1 are in order. First, as the seriousness of not meeting several perpetual obligations rises to very high to extreme levels, the level of those risks morphs into extinction risks. For example, if involuntary environmental risks become extreme, logically the magnitude of the risks could reach the level of human extinction. Thus, the entire analytical and action framework developed around human extinction would then come into play. This same logic applies to four other perpetual obligations: bequeath sustainable societies; bequeath sustainable production; protect the essence of nature; and protect the essence of humans.

Second, the maintaining options perpetual obligation is an outlier because the more extreme components of the action framework are contradictory in spirit to meeting this perpetual obligation. Certainly, ubiquitous command and control measures directly act to limit behavioral options at a large scale. Additionally, in very serious situations, such as a war footing situation with respect to human extinction, promoting the maintaining options perpetual obligation could be exceptionally counterproductive and simply not justifiable.

Third, some action framework components are not applicable to some perpetual obligations. This is the case with respect to three of the last four perpetual obligations. For example, it seems inconceivable to devote 15% of GDP or even 5% of GDP to fully inform the world's citizens about futures. Likewise, those levels of funding seem to be inappropriately high to preserve the humanity's knowledge and even for research and development programs.

Next, I need to say a few words about the unfinished business obligation and a couple of the action categories. As noted in Chapter 6, we should regret as a species to not finish our unfinished business but maybe at this point in time, we may not know what our unfinished business actually is. As a baseline, I am assuming our unfinished business may relate to major humanity-wide projects such as building WEMS or a global electricity grid, which would require a Major Programs level of effort for many years. However, it seems plausible that extraordinary possibilities might present themselves to humanity, maybe once in every million years or even once in every billion years that would require Command and Controls and War Footing action levels, respectively (see Table 7.1 for these entries). Again, I cannot profess to knowing what these opportunities might be at this point in time, but it makes sense to allow for such opportunities in the development of the action framework.

Lastly, it is very reasonable to expect that efforts to reduce the extent we are not meeting each perpetual obligation could be combined into more comprehensive efforts. In this manner, separate Manhattan Project levels of effort could be combined to still fall within 5% of global GDP instead of accumulating to a much higher level of effort. This is illustrated more fully in the following fourth section. Additionally, the mapping of the extent to which we are not meeting each obligation to each action category is presented as a guideline, not as immutable. For example, there may be instances where a perpetual obligation is not being met to a large degree but that the solutions only require policy nudges.

Dealing with uncertainties: human extinction example

This section addresses the topic of uncertainty. The basic assumption is that more intensive and pervasive actions will be needed the more that a perpetual obligation is not being met. In the case of human extinction, for example, more intensive and pervasive actions will be needed the more that the probability of human extinction exceeds the 10^{-20} threshold. Of course, there will be substantial uncertainties related to the metrics associated with each of the perpetual obligations.

The standard approach to express uncertainties is by using probabilities. For example, the chance of rain tomorrow could be 50%. The odds of your favorite sports team winning this year's championship could be 5 to 1. The probability of winning the lottery could be 1 in 10 million. We have already used probabilities as metrics for thresholds for the human and earth life extinction perpetual obligations.

Uncertainties associated with assessing perpetual obligations are qualitatively different from estimating the chance of rain over the next couple of days. The uncertainties are larger, and our knowledge is more limited. In these cases, I prefer, and recommend here, to use lower and upper probabilities to represent what is known about meeting each perpetual obligation.[7] Technically, lower and upper probabilities fall within a branch of mathematics known as imprecise probabilities.[8] Imprecise probabilities relax a few key assumptions that are found in more classical expositions of probability.

What appeals to me about upper and lower probabilities is that one can represent what is known or not known and represent improvements in knowledge. Here is a thought experiment. Consider an urn with blue and red balls. We know there are 100 balls in the urn. The challenge is to estimate the probability of drawing a blue ball out of the urn. If we knew the exact color of every ball, this task is very straightforward. Unfortunately, in this case, we do not know the exact color of every ball. We can only see a few balls from the outside. We can see that 20 balls are blue and 20 are red.

Using imprecise probabilities, we would estimate the lower probability of drawing a blue ball as 20% and the upper probability as 80%, if all of the unseen balls are red or blue, respectively. Using a classical probability approach, one might be tempted to say that the probability of drawing a blue ball is simply 50%, which to me seems to imply that we have more knowledge about the situation in the urn than we actually do.

To better understand this, let's now assume that we find a way to see more balls in the urn, such as by uncovering the top and the bottom of the urn. We can now see 50 balls and know that at least 25 are blue and 25 red. Applying the imprecise probability approach, now we know that the lower probability of drawing a blue ball as 25% and the upper probability as 75%, given that all of the unseen balls could be red or blue, respectively. As our knowledge increases, the range between the upper and lower probabilities decreases. The probability would still be 50% using the classical probability approach, which does not reflect a reduction in our ignorance about the number of balls in the urn.[9]

Within the world of imprecise probabilities, I see two approaches that can be of value to dealing with the substantial uncertainties associated with judging whether we are meeting our perpetual obligations. The first approach is most straightforward and entails directly estimating upper and lower probabilities. This approach is most straightforwardly applied to the human extinction, species extinction, and involuntary environmentally related risk of death perpetual obligations whose metrics are actually stated as probabilities. The basic approach is to estimate upper and lower probabilities and then assess which category of action is more appropriate.

Figure 7.2 illustrates how pairs of upper and lower probabilities could map to each of the six action categories introduced earlier with respect to the perpetual obligation to prevent human extinction. The dark lines indicate the upper probabilities and the dashed lines indicate the lower probabilities. The dotted line is the 10^{-20} ethical threshold. The lower and upper probabilities are shown to stay the same over the 1000-year planning period just for convenience. For example, in the Do Nothing case, dark and dashed lines are shown significantly below the threshold and in the War Footing case the two lines are close together and are shown very significantly above the threshold. Only Policy Nudges might be needed if our ignorance about the probability of human extinction is relatively low (i.e., the upper and lower probability lines are close to each other), and these upper and lower probabilities are close to the threshold.

The lower and upper probabilities could be curvilinear over time in very complicated patterns, which could then pose intellectually challenging situations to determine what should be done when. Some examples of challenging situations are depicted in Figure 7.3. Should humanity be less active in reducing the probability of extinction, which exceeds the threshold in the short-to-mid term but decreases to acceptable levels in the longer term, as depicted in Case D? What are we to make of the curves in the second case, which suggest that our knowledge about the probability of human extinction severely drops off over time while ironically indicating that the lower probability of extinction could be quite low? Maybe in this case, humanity is not meeting Perpetual Obligation #10, Support the Generation of New Knowledge, very well at all and a Manhattan-scale effort is needed to expeditiously reduce our ignorance.

Approaches will be needed to decide how to interpret upper and lower probabilities that provide mixed messages. For example, does the Case F in Figure 7.3 suggest that action is needed, since the preponderance of the upper probability curve exceeds the ethical threshold, or not, since the lower probability curve does not exceed the threshold? What does risk adverse mean in this context? How might the Precautionary Principle be applied?[10] To what degree does the concept of regret impact decision-making this realm? It will likely be the case that these approaches will emerge organically from judgments made by the anticipatory institutions introduced and described in the next chapter.

The second way of employing imprecise probabilities is through evidential reasoning. Imagine that you are a juror and are responsible for judging whether a defendant is innocent or guilty. The prosecutors and defense attorneys have presented to you numerous pieces of evidence that support

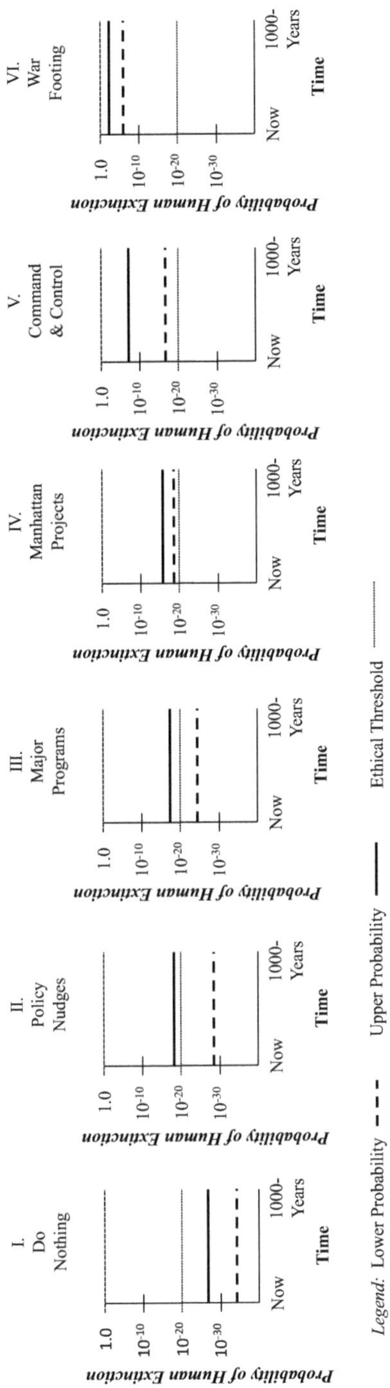

Figure 7.2 Applying Lower and Upper Probabilities to the Global Action Framework: Human Extinction Example

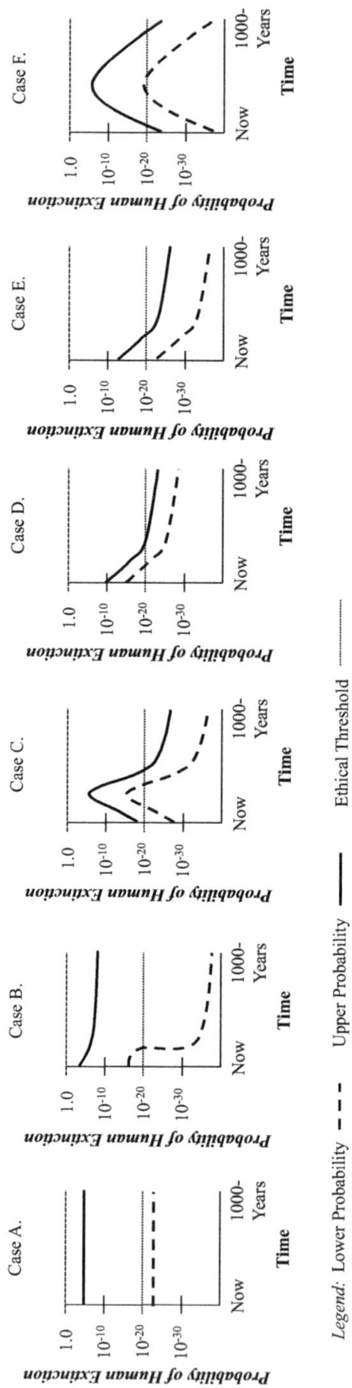

Figure 7.3 Ethically Challenging Lower and Upper Probabilities with Respect to Human Extinction

one verdict or the other and maybe even in some cases, both verdicts. You then carefully weigh evidence to render your verdict.

Evidential reasoning with imprecise probabilities is quite similar. In this case, the set of judgments could be greater than two. In our case, the set of potential judgments contains six options, action categories I through VI. Evidence is gathered that pertains to making an action category judgment. Imprecise probabilities can be assigned to single action categories but also can be assigned to subsets of action categories when one is uncertain about how to allocate probabilities to specific action categories. Imagine that you are a physician and have conducted several tests to determine a patient's diagnosis. One test result may suggest one of three life-threatening diagnoses but cannot pinpoint the exact diagnosis. Another test may weakly support one of these three diagnoses and a fourth serious diagnosis. The results of a third, potentially conclusive, test may take a week to receive. What should the physician do in the meantime? They may elect to pursue supportive care and treat for three diagnoses, at least in the short term and as long as the treatments were not contradictory.

Quantitative approaches have been developed to assign imprecise probabilities to sets of potential diagnoses or judgments based on individual pieces of evidence and then to combine these probabilities to suggest which diagnosis has the most support.[11] At a minimum, the evidential reasoning paradigm should be applied qualitatively to assess perpetual obligations with metrics that are not expressed as probabilities. Let's use Perpetual Obligation #3 as an example. What pieces of evidence might one assemble to judge to what extent humanity is meeting this perpetual obligation? In Chapter 6, it was recommended that one could rely upon existing freedom, democracy, and human rights indices and also make use of the global set of SCES. One could also survey the current political landscape and assess the most influential trends. When I consider these types of pieces of evidence, my judgment is that the requisite action level is a major program. Of course, my judgment could be made more rigorous and transparent if I took the time to explicitly identify and describe the pieces of evidence I am using, assign lower and upper probabilities for each piece of evidence to all subsets of actions, and then applying an algorithm to combine these assessments to reveal which action accrued the most evidential weight.

Meeting perpetual obligations

Table 7.2 presents my judgments about what levels of actions need to be taken by humanity with respect to each perpetual obligation based on situations circa the year 2020. The actions represent a mix of Level II, III, and IV actions to prevent human and earth life extinction, bequeathing sustainable systems of production and reducing involuntary environmental risks. These recommended actions are in contrast to a mixture of Do Nothing

Table 7.2 Assessment of Levels of Global Actions Facing Humanity

Perpetual Obligations	*Suggested Action Level*	*Comments*
#1. *Prevent human extinction* #2. *Prevent extinction of life on earth* #4. *Bequeath sustainable production* #5. *Reduce involuntary environmental risks* #10. *Support generation of new knowledge* #11. *Finish important business*	Manhattan Project	Contributory projects would address clean and renewable energy and manufacturing processes. Funds would also be allocated to develop WEMS, GADSS, and the global electricity grid.
#3. Bequeath sustainable societies	Major Program	This program would be composed of citizen education and engagement processes, enlightened improvements of social media, and activities to increase racial and ethnic tolerance.
#6. Preserve essence of nature #7. Preserve essence of humans #8. Maintain options #9. Preserve human knowledge #12. Fully inform current generations	Policy Nudges	Policies would focus on better oversight of genetic engineering, improving support for the arts and humanities, and increasing futures literacy. A separate effort would focus on ensuring that human knowledge is being preserved.

and Policy Nudges policies that dominate current actions related to these perpetual obligations.

Specifically, I currently believe that the lower and upper probabilities with respect to preventing human extinction exceed ethical thresholds and therefore are unacceptable. The Manhattan Projects graph in Figure 7.2 most closely resembles my judgments about the lower and upper probabilities of human extinction. Conversely, I believe that at least 1 in a 1000 SCES that results in human extinction could also result in the extinction of earth life (see Exhibit 6.2 for a graphical presentation of ELExS). Both of these judgments fall into the Manhattan Projects action level indicated in Table 7.2.

I also believe that our current systems of production are extremely unsustainable. The question for me is whether production systems are on a course to become more sustainable and if so, quickly enough to make a substantial impact on our ability to meet Perpetual Obligation #4. The evidence is decidedly mixed. We are seeing that the energy intensities of national economies are at least holding steady even as economies expand. We are also seeing increases in the use of renewable and clean energy. On the other hand, greenhouse gas emissions continue to increase. Air and water pollution levels are still too high, as are the use of pesticides and herbicides. In a thousand years, at the point of the Singularity, humanity might achieve this goal. Unfortunately, the risks posed in the interim contribute to issues with respect to Perpetual Obligations #1 and 2. Thus, I also judge the extent of not meeting this perpetual obligation as falling into the Manhattan Projects category. Because of our environmentally unsustainable production systems and risks to individuals from climate change, I also believe that the level of involuntary environmental risks faced by individuals across the world is much higher than the 10^{-6} threshold (see Figure 6.4), and therefore also require Manhattan-Scale projects to rectify.

I believe that issues with respect to these four perpetual obligations are inextricably linked and can be addressed by one overarching Manhattan Project effort. I also believe that significant research must be done to complement these activities (Perpetual Obligation #10) and that these efforts could also benefit from the contributions of unfinished business projects (Perpetual Obligation #11). Here I am thinking about WEMS, a global electricity grid, and GADSS, which is introduced later in this chapter. Coalescing efforts that focus on these six perpetual obligations could be managed with a Manhattan Project scale effort in the 5% of global GDP range.

I am not as worried about failures to protect the essence of nature and humans (Perpetual Obligations #6 and #7) at this point, as genetic engineering of nature and humans is still in its infancy. We have time to enact Policy Nudges to preclude serious threats to meeting these perpetual obligations. My current observations about meeting the maintaining options perpetual obligation is that Policy Nudges are needed to improve freedoms across the world and increase funding for creative endeavors in the arts and humanities, but that for the most part cultures are still vibrant and evolving. I think the level of action represented by Policy Nudges is also justifiable with respect to Perpetual Obligations #9 Preserve Human Knowledge and #12 Fully Inform Current Generations. With respect to the former, it could be that straightforward policies could be instituted to more fully safeguard human knowledge that is accumulating in the cloud. With respect to the latter, policy nudges across the globe could bring futures literacy training into educational curricula for all age levels.

A major program level of action is recommended with respect to Perpetual Obligation #3, Bequeath Sustainable Societies. In my opinion, major social justice issues have not been resolved and may be worsening. To blame are corrosive impacts of social media, inflammatory politics, and painful economic uncertainties inflicted on households across the world by globalization, national economic crises, and climate change.

To reiterate, these are my own judgments, earnestly given with the additional purpose of illustrating how humanity can operationalize in a policy context meeting perpetual obligations. In the next chapter, it is recommended that an international organization and national anticipatory institutions formally implement processes to make these judgments.

1000-year planning

Regardless of the level of global actions facing humanity, humanity ought to undertake very long-term planning efforts. Let's refer to this activity as 1000-year planning. This span matches the perpetual obligation assessment time frame portrayed earlier in Figures 7.2 and 7.3. This time frame also allows humanity to gain new insights into the challenges we need to face and the solutions open to us. For example, not in the next 50 years but sometime during the next 1000 years, we will most assuredly exhaust our fossil fuels if we do not curb our use. The potential collapse of the global energy sector could call into question the ability of current generations to meet several perpetual obligations, including the prevention of human extinction and maintenance of options obligations. Certainly, progress is being made to build more renewables into our energy infrastructure, but near-term economic calculations constantly constrain this progress. Additionally, myopic policy perspectives are also constraining investments in transformative science and technology that could solve our energy needs for many thousands of years, but which could take centuries to perfect. Here I am thinking about the prospects for fusion energy, space-based microwave systems, and a global grid that would link areas of the earth that have low populations but very high renewable energy resources to areas with high population densities but low renewable energy resource potential.

Energy is not the only topic amenable to 1000-year planning. Previously, we noted that wastes from nuclear power plants could be toxic for approximately 10,000 years. At least in the United States, the inability for political processes to adopt long-term time frames means that waste is accumulating in canisters at the sites of nuclear reactors, which is definitely not a long-term solution. Proposals to establish a long-term nuclear waste repository, such as at the Yucca Mountain site, are on indefinite hold. It would be preferable to have a long-term plan in place to deal with nuclear wastes.

In another vein, humanity's exploration of space, which surely could encompass inspiring unfinished business projects, is hobbled by short-term economic and political concerns. Our current presence in and exploration of space is miniscule, especially compared to humanity's military investments. No long-term vision that could span 1000 years has been articulated. Given the technical challenges and the sheer amount of time needed to explore space, this topic seems to be a natural for a 1000-year planning effort.

Thousand-year planning is needed with respect to the climate crisis. The climate crisis is already resulting in the loss of homes and forced migration. As the situation worsens and families are forced out of their homes due to sea-level rise, inland flooding, and desertification, new homes will need to be found somewhere on the earth. Global human settlement planning that transcends national borders will be needed to reduce human suffering and the potential for international conflict. Housing is only one aspect of this challenge. Every other aspect of urban infrastructures will also need to be addressed, from transportation to drinking water to electricity. One could even imagine city-level infrastructure and comprehensive land use plans spanning 1000-years. All of these challenges could be better met if humanity were investing more resources to meet the spirit of Perpetual Obligation #10, supporting the generation of new knowledge to ensure the journey and meet the other perpetual obligations.

Figure 7.4 presents a schematic to guide humanity's 1000-year planning efforts. The schematic plots topics by planning time frames, as not all topics can or ought to be considered within the full 1000-year time frame. It should also be stressed that 1000-year plans ought to be reevaluated regularly and frequently, for fine-tuning or major revisions as situations warrant. On the other hand, we do not want the plans to be open for alteration willy-nilly at the whim of short-term, myopic, and politically motivated decision-making. For example, a 1000-year program to explore space needs to have a firm vision to guide long-term R&D and investments in infrastructure. Exploration goals need to be steady. However, the best laid plans could change given breathtaking technological breakthroughs, newly realized opportunities, or even newly detected extraterrestrial threats.

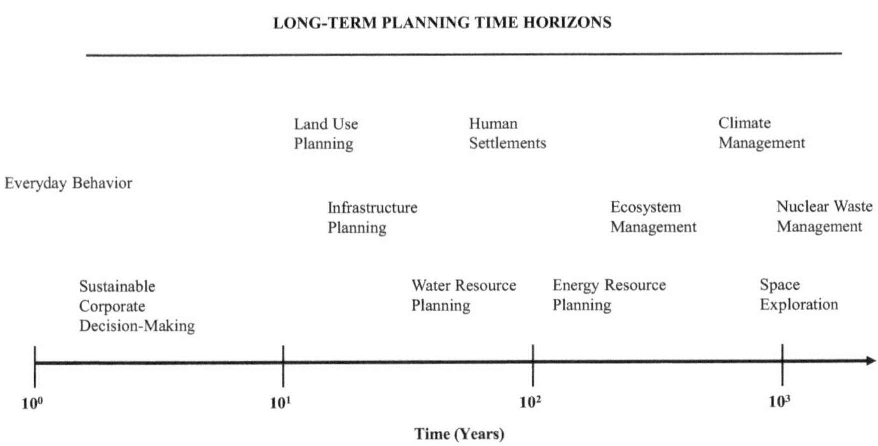

Figure 7.4 Schematic for Global 1000-Year Planning

Integrating global goals and sub-global actions

Now it is time to move from perpetual obligations that pertain to generations across all time and space to goals and actions that are approachable and achievable by current generations. The challenge is to move top-down first to global goals and then to sub-global actions. In other words, there is the need to move down to actions that nations, states, and localities can implement that have dual purposes: meet the needs of current generations and also meet perpetual obligations. It is also important that actions synchronize with each other.

The United Nations Sustainable Development Goals (SDGs) present an excellent example of global goals that fill the space between perpetual obligations and national and subnational actions. The 17 SDGs can be found in Exhibit 7.1. The preponderance of the SDGs can be viewed as strategies related to two perpetual obligations, bequeath sustainable societies (#3) and bequeath sustainable systems of production (#4). Reduced inequality and quality education are examples that relate to the former and affordable and clean energy and responsible consumption and production relate to the latter. Meeting the SDGs will vastly improve the lives of billions of people and will contribute to meeting a small number of perpetual obligations but in and of themselves are not biggest picture possible ideas. Additionally, national and subnational actions are needed to meet SDGs.

Other sources of global goals that sit in between perpetual obligations include:

- Global goals for GHG reductions stipulated in agreements such as the Paris Agreement and the Kyoto Protocol
- Global GHG emission reduction solutions put forth by Project Drawdown[12]
- International goals with respect to space exploration[13]
- World Health Organization goals that provide additional details with respect to SDG #3 (e.g., WHO subgoal 3.1 to reduce global maternal mortality ratio to less than 70 per 100,000 live births)[14]
- Species protection goals set out in the Convention on Biodiversity[15]
- International statement on human gene editing[16]

Exhibit 7.1 United Nations Sustainable Development Goals (SDGs)*

- Goal 1. End poverty in all its forms everywhere
- Goal 2. End hunger, achieve food security and improved nutrition, and promote sustainable agriculture

- Goal 3. Ensure healthy lives and promote well-being for all at all ages
- Goal 4. Ensure inclusive and equitable quality education and promote lifelong learning opportunities for all
- Goal 5. Achieve gender equality and empower all women and girls
- Goal 6. Ensure availability and sustainable management of water and sanitation for all
- Goal 7. Ensure access to affordable, reliable, sustainable, and modern energy for all
- Goal 8. Promote sustained, inclusive, and sustainable economic growth, full and productive employment and decent work for all
- Goal 9. Build resilient infrastructure, promote inclusive and sustainable industrialization, and foster innovation
- Goal 10. Reduce inequality within and among countries
- Goal 11. Make cities and human settlements inclusive, safe, resilient, and sustainable
- Goal 12. Ensure sustainable consumption and production patterns
- Goal 13. Take urgent action to combat climate change and its impacts*
- Goal 14. Conserve and sustainably use the oceans, seas, and marine resources for sustainable development
- Goal 15. Protect, restore, and promote sustainable use of terrestrial ecosystems, sustainably manage forests, combat desertification, and halt and reverse land degradation and halt biodiversity loss
- Goal 16. Promote peaceful and inclusive societies for sustainable development, provide access to justice for all and build effective, accountable, and inclusive institutions at all levels
- Goal 17. Strengthen the means of implementation and revitalize the global partnership for sustainable development

* https://sustainabledevelopment.un/?menu=1300

Let's next link perpetual obligations to global goals and then to national and subnational actions. Figure 7.5 presents an example related to climate change and transportation decision-making. At the level of perpetual obligations, climate change is an existential threat to humanity and earth life, increases involuntary environmental risks (e.g., from extreme weather events), and arises in large part from unsustainable production systems. International climate agreements and goals contained therein can help meet these perpetual obligations. Also, numerous SDGs pertain to both climate change and transportation, such as SDG # 13, 7, 9, 11, and 15.

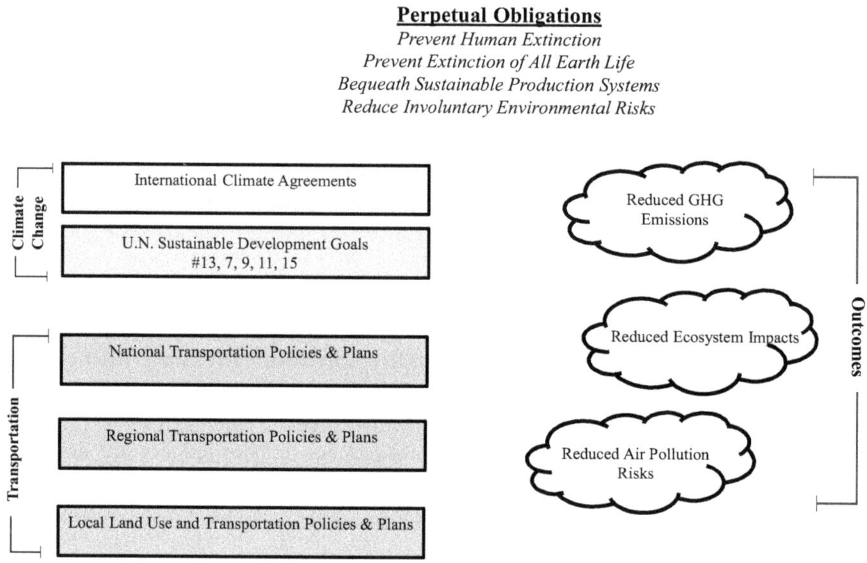

Figure 7.5 Linking Perpetual Obligations, Climate Change, and Transportation Policy

Nations can focus on transportation policies and plans, among many pathways to help meet goals set out by the SDGs and international climate agreements. Reducing fossil fuel use by the transportation sector can be accomplished through increases in fuel efficiency and market penetration of electric vehicles, assuming that electricity generated for the EVs emits no or low levels of GHGs. Decisions can also be made at regional levels with respect to rail and mass transit services. Then, many important decisions are made at the local level that can reduce travel demands by fossil fueled vehicles and vehicle miles traveled. Inextricably tied to transportation planning at the local level is land use planning. In the United States, land use and transportation planning decisions have fostered sprawl, which greatly increased GHG emissions from vehicles and is a major constraint to reducing GHG emissions moving forward.

In the United States, it must be noted that for the most part, local land use decision-making is wholly divorced from federal programs that aim to reduce GHG emissions. In other words, the federal government does not dictate local land use planning decisions, so there is always a regulatory valley between energy and environmental policy making in the United States. The gap needs to be overcome if fully integrated goals and actions that can all effectively contribute to meeting perpetual obligations. The aspirational goal is a world where local land use decisions are made not only with local issues in mind but also with perpetual obligations in mind, too.

Similar multilevel collaborative efforts are needed for linking perpetual obligations with national and subnational efforts. However, the models may differ from the one presented in Figure 7.5. A model for futures literacy and communication, Perpetual Obligation #12, could look fairly similar in that nations, states, and localities could all have futures literacy plans and policies. A model of plans and policies with respect to CRISPR and protecting the essence of humanity might include national policies but would then need to include research funders as well as universities and the private sector. A model with respect to finishing unfinished business might only involve international collaborations on such projects with contributions made by nation states.

Global anticipation and decision support system

A secondary theme is arising in this book and that is the use of the broadest range of information technologies integrated worldwide to support and ensure the journey and meeting our perpetual obligations. The first set of recommendations was introduced in Chapter 4 and pertained to the Whole Earth Monitoring System. Here we propose an integrated system to support anticipation and long-term planning with respect to the dozen perpetual obligations and decision-making across the globe at all scales of government and society that focuses on goals and interests of both current generation and future generation. Let's call this system the Global Anticipation and Decision Support System (GADSS). Like WEMS, building GADSS can be considered a capstone project of humanity.[17]

Figure 7.6 presents a schematic of GADSS. The major inputs into GADSS are the perpetual obligations metrics (e.g., upper and lower probability curves; evidence pertinent to metrics for qualitative obligations) and suggested action levels. Based on these inputs, stakeholders (which include all levels of government, private and non-profit organizations, the noosphere[18] for decision makers) develop plans and policies, which contribute to meeting humanity's perpetual obligations as well as meeting specific stakeholders' needs. GADSS aggregates the results and feeds the results back to the stakeholders. In this way, stakeholders can learn what to anticipate with respect to human behavior, not just with respect to larger forces of nature. It is anticipated that this process will also foster cooperation and collaboration as well. This feedback process can run iteratively based on data until an equilibrium of plans and policies is found. Evolving data may mean that metrics should be reconsidered. In the extreme, reflection may result in revised definitions of perpetual obligations.

GADSS is conceptualized as being comprehensively systems oriented. It should be designed to inform users about how the system of systems operates and how and what they can anticipate with respect to positive, negative, and non-linear systems feedback. Improving systems literacy worldwide could have profound positive impacts on all levels of decision-making.

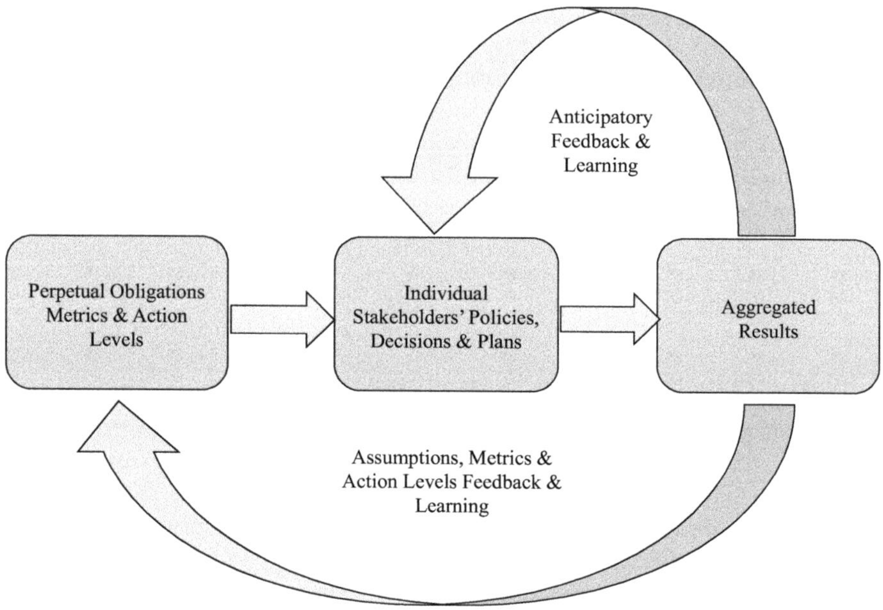

Figure 7.6 Schematic for the Global Anticipation and Decision Support System

Here I turn to the work of Peter Senge and his book, *The Fifth Discipline*, on systems thinking.[19] Not only is it important to understand the unintended consequences of decision-making that does not fully appreciate and understand positive and negative feedback loops (e.g., as Senge illustrates by the Beer Game, where demands and supplies of beer can be widely distorted by a lack of systems perspectives), but it is also important to develop common mindsets about systems, feedback, and anticipation. Stove-piped decision makers isolated from anticipatory inputs and feedback from the system of systems will not develop a common understanding of the world. They will behave antagonistically against each other, however unwittingly. GADSS, on the other hand, will draw everyone in. Everyone will get feedback. Everyone will see how the system of systems operates and will understand both positive and negative feedback effects. Maybe this higher understanding will also help to raise consciousness about the journey and perpetual obligations.

GADSS would be designed using agent-based modeling (ABM) concepts. In this case, each agent would map to a specific and actual decision maker, such as the office of a land use planner in a major urban area. The software agents would not have to be imbued with decision-making rules, as the decision makers themselves would decide what to do given all relevant inputs tracked by the system to that point. The ABM component of GADSS would then play out the impacts of the decisions on each of the agents (i.e., the purviews of each decision-making unit). Numerous rounds of exploratory decision-making could be held to allow participants to anticipate outcomes from the system of decisions and to eventually post commitments to decisions.

There are many organizations around the world that already serve as coordination points. The World Health Organization and the Organisation for Economic Co-operation and Development come to mind. GADSS can assist public health officials across the world by helping them anticipate the impacts of decisions made in their own nation or on the other side of the planet.[20] GADSS would unite anticipation and planning efforts to support synergistically balanced decision-making. GADSS would be able to share policy intentions, say with respect to land use planning, with users and would also support the gaming of decisions.

GADSS would generate various scorecards based on meta-assessments of the results of each gaming session. These would include the scorecards already presented earlier, including the existential risk scorecard presented in Chapter 3, the meeting perpetual obligations scorecard from Chapter 6, and the global action to meet perpetual obligations scorecard presented earlier in this chapter.

GADSS would also be designed to assist users better comprehend system of systems dynamics and understand and detect emergent phenomena. In these ways, GADSS could assist the world of decision makers better understand the unintended consequences of their decisions as well as help them better deal with unknown unknowns.

One could imagine that the design would have these types of intelligent system capabilities[21]:

- Communicate results of important anticipatory exercises;
- Help users create problem statements and build simple decision matrices;
- Connect users seamlessly to WEMS;
- Automatically couple models and run models to meet user needs;
- Modify user interfaces to better communicate complex information and to overcome common heuristics and biases that afflict human decision-making[22] and understanding and processing uncertainty estimates;
- Monitor, distill, and synthesize users' comments for everyone's use;
- Learn individual user values and proposes policy options with evaluation criteria;
- Simultaneously interact with multitudes of users to learn their values with respect to policy options;
- Develop proposed policy options to satisfy values expressed by multitudes of users;
- Facilitate discussion, negotiation, and mediation with respect to policy decisions;
- Assist users in evaluating past decisions and policies.

Upon reflection, the scale and scope of GADSS and WEMS combined probably represent an example of an unfinished business project for humanity. Technology does not yet exist to provide many of the capabilities mentioned here. Additionally, the level of coordination and trust needed across the world to build these systems is unprecedented. Certainly, extraordinarily high levels of coordination and trust will be necessary for humanity to survive deep into time and space, and a GADSS system could lay the foundations.

Notes

1 This and the following two sections are adapted from B. E. Tonn. 2017. Philosophical, Institutional, and Decision Making Framework for Meeting Obligations to Future Generations. *Futures*, 95, 44–57.
2 T. Friedman. 2004. A New Mission for America. *The International Herald Tribune.*
3 D. Stine. 2009. *The Manhattan Project, Apollo Program, and Federal Energy Technology R&D Programs: A Comparative Analysis.* Congressional Research Service, Washington, DC.
4 J. S. Bardi. 2008. *Stephen Hawking Renews Call to Colonize Space.* American Institute of Physics, College Park, MD; S. D. Baum. 2010. Is Humanity Doomed? Insights From Astrobiology. *Sustainability*, 2, 591–603; www.telegraph.co.uk/science/2017/06/20/human-race-doomed-do-not-colonise-moon-mars-says-stephen-hawking/
5 E. Kintisch. 2010. *Hack the Planet: Science's Best Hope – or Worst Nightmare – for Averting Climate Catastrophe.* John Wiley & Sons, Hoboken, NJ.
6 C. Tassava. 2010. The American Economy during World War II. *EH.net.* Retrieved from https://eh.net/encyclopedia/the-american-economy-during-world-war-ii/
7 Other researchers might advocate using a Bayesian approach to addressing uncertainties in this context. For an interesting comparison between the two approaches, see G. Shafer and A. Tversky. 1985. Languages and Designs for Probability Judgment. *Cognitive Science*, 9, 3, 309–339.
8 G. Shafer. 1978. Non-Additive Probabilities in the Work of Bernoulli and Lambert. *Archive for History of Exact Sciences*, 19, 4, 309–370; Peter Walley. 1991. *Statistical Reasoning with Imprecise Probabilities.* Chapman and Hall, London; A. P. Dempster. 1967. Upper and Lower Probabilities Induced by a Multivalued Mapping. *The Annals of Mathematical Statistics*, 38, 2, 325–339.
9 I realize that is an exceedingly shallow discussion of classical and imprecise probabilities. The goal of this book is to set out a blueprint, which I believe should include the use of imprecise probabilities. An in-depth, technical debate about the merits of imprecise and classical probabilities for use in the context of this book is an excellent topic for future work. It should also be noted that due to inherent uncertainties that are characteristic of non-linear and chaotic systems, that it is probably unreasonable to expect upper and lower probability curves to ever coalesce into classical probability distributions.
10 D. Kriebel, J. Tickner, P. Epstein, J. Lemons, R. Levins, E. Loechler, M. Quinn, R. Rudel, T. Schettler, and M. Stoto. 2001. The Precautionary Principle in Environmental Science. *Environmental Health Perspectives*, 109, 9, 871–876.
11 G. Shafer. 1976. *A Mathematical Theory of Evidence.* Princeton University Press, Princeton, NJ.
12 P. Hawken (Ed.). 2017. *Drawdown: The Most Comprehensive Plan Ever Proposed to Reverse Global Warming.* Penguin Books, New York.

13 www.globalspaceexploration.org/wordpress/wp-content/uploads/IAC62/ISECG%20Goals%20Obj%20Benefits_IAC-11-A5.4.2_Paper.pdf

14 www.who.int/topics/sustainable-development-goals/targets/en/

15 www.cbd.int/doc/c/efb0/1f84/a892b98d2982a829962b6371/wg2020-02-03-en.pdf

16 www.nationalacademies.org/news/2015/12/on-human-gene-editing-international-summit-statement

17 Obviously, I believe in advanced information technologies. I am old enough to have used punch cards to program computers as an undergraduate at Stanford. Being a Stanford grad, I have an affinity to Silicon Valley and its start-up culture, especially in the IT space. At Harvard as a graduate student in the late 1970s, I was a Fortran programmer at the Laboratory for Computer Graphics and Spatial Analysis. At ORNL, I also had the opportunity to participate in the field of artificial intelligence. I remember trying to program in LISP on our Symbolics machine. I remember meeting Dr. John Holland, the inventor of the genetic algorithm. However, the experience that has had the most lasting influence came during a conference about Computing for the Social Sciences held in 1993. A young researcher from the University of Illinois-Champaign-Urbana named Marc Andreessen made a presentation on a piece of software he was working on, a beta version of Mosaic. I remember telling my colleagues back at ORNL about Mosaic and how it linked computer sites together (i.e., a web browser). At the time, I thought it was too bad there weren't more computers to link to around the world. Mosaic became Netscape, which was followed by Internet Explorer and other web browsers. Then Google debuted in 1995 after being hatched at Stanford. Then the web exploded. I find this all astounding and hope that WEMS and GADSS will be, too.

18 Here I am using the term noosphere to the space of shared consciousness among the world's decision makers. For more on this concept, see https://en.wikipedia.org/wiki/Noosphere

19 P. Senge. 1990. *The Fifth Discipline: The Art & Practice of the Learning Organization.* Currency/Doubleday, New York.

20 Imagine if GADDS had been in existence at the beginning of the coronavirus pandemic and if it had fostered seamless international cooperation to contain its spread.

21 B. E. Tonn and D. Stiefel. 2012. The Future of Governance and the Use of Advanced Information Technologies. *Futures*, 44, 812–822.

22 D. Kahneman, P. Slovic, and A. Tversky (Eds.). 1982. *Judgement Under Uncertainty: Heuristics and Biases.* Cambridge University Press, New York.

8 Anticipatory institutions and perpetual organizations

This chapter tackles institutional innovations needed to support the program set out in the previous three chapters. A robust framework of futures-oriented institutions can help build individuals' confidence that their caring for future generations could influence policy. This framework will also be responsible for judging whether humanity is meeting its perpetual obligations and if not, what actions should humanity consider to rectify the situation.

The focal point of the institutional innovations presented in this chapter is national anticipatory institutions (NAIs). The first section posits that these new institutions would judge from their own national perspectives about whether humanity is meeting its perpetual obligations. These institutions would be created through amendments to national constitutions. The World Court of Generations (WCG) would represent all of humanity when it also renders a judgment on this question. The Intergovernmental Panel on Perpetual Obligations (IPPO) would be responsible, in part, for providing the NAIs and WCG with information gleaned from WEMS and GADSS with metrics discussed in Chapter 7 to help with this task.

The goal of this complicated decision-making process is the development of an international agreement, a 1000-year plan if you will, on how to meet perpetual obligations, which is referred to as the Perpetual Obligations International Protocol (POIP). The chapter concludes with a discussion on the need for a small number of perpetual organizations that will be able to serve humanity throughout times of extreme turmoil.

National anticipatory institutions

It is assumed that governments across the globe, even those that are democratic, are too myopic, too inflicted with presentism, to adequately represent the interests of future generations.[1] New national institutions, then, are needed to fill this role. The main responsibility of NAIs will be to engage their nation's citizens in a process to assess whether their current generations are meeting their perpetual obligations. My conception of the NAIs is that they will be the care-for-future-generations' conscience for their nations. The NAIs are intended to be the angel in the ears of national leaders and citizens. They are intended to be granted the gravitas needed to bolster the credibility of their assessments. They are not intended to be embroiled in everyday politics or in decisions about actual policies and expenditure of funds pledged by national governments. Thus, NAIs are envisioned to be institutions separate from legislative bodies as much as possible in order to reduce the potential for the corruption of their role as angels in the ears who advocate for future generations.

It is easy for countries to convene commissions and panels to take on this responsibility. Indeed, a report prepared by the Institute for European Environmental Policy identifies several organizations that have been founded around the world to take on these responsibilities.[2] These include the Welsh Commissioner for Future Generations and the Canadian Commissioner of the Environment and Sustainable Development. Finland, Germany, Israel, Malta, United Kingdom, and Sweden have or have had futures-oriented organizations that serve at the national level. The report goes on to recommend that establishment of a European Union Guardian for Future Generations.[3]

An issue with these organizations is that they can be easily defunded and ignored. Therefore, it is recommended that every nation formally stipulates in their constitution a provision that addresses responsibilities for advocating for future generations and the journey. Explicit stipulation of obligations to future generations in constitutions may provide the legitimacy and gravitas needed for this task. The next section of this chapter provides the results of an assessment of 100 national constitutions to see to what extent, if any, national constitutions already have explicit provisions for NAIs and to what extent, if at all, national constitutions address futures and future generations.

Lack of national anticipatory institutions

The resources of the Constitute Project were used to access the most recent English language versions of 100 national constitutions.[4] Combined, the 100 countries account for 90% of the world's population and include the top 40 most populous countries. The distribution of countries across continents is Africa 17%; Asia 20%; the Americas 14%; Europe 27%; Middle East 11%; and Oceania and Island States 11%. I believe that the 100 countries are representative with respect to income, religion, political structure, and freedom (see Table 8.3 for additional descriptive statistics about the sample of countries and their constitutions).

Each constitution was carefully reviewed for provisions that explicitly assign responsibilities for advocating for future generations to specific institutions, such as commissions or councils. Additionally, I conducted a keyword search of the 100 constitutions, searching for the words: future generation(s); posterity; anticipation; and foresight. The key word search would reveal said institutions but also provide a more general perspective on how futures-oriented a constitution might be.

I assessed the futures-orientation of each constitution using this rubric:

- No use of key terms
- One or two uses of key terms
- Several uses of key terms
- Numerous uses of key terms
- One explicit provision for an institution to advocate for future generations
- Two or more explicit provisions for institutions to advocate for future generations

Of the 100 reviewed constitutions, only two contained explicit provisions for institutions to advocate for future generations:

- Hungary – The Commissioner for Fundamental Rights has the responsibility for protecting future generations.
- Tunisia – The Commission for Sustainable Development and the Rights of Future Generations has this responsibility.

As indicated in Table 8.1, just over 50% of the reviewed constitutions contained no provisions for institutions to advocate for future generations or any of the key words. In other words, futures or even posterity were not concepts that frequently made it into the constitutions. These constitutions focused solely on setting out the nuts and bolts of governance: the legislature, the executive branch, the judiciary. When key words were used, they were most commonly found in preambles and in references to commitments to sustainable development.[5] Overall, I am not surprised at the lack of formal institutions related to future generations specified in constitutions. But I am surprised at the lack of vision or even the notion of future generations and posterity.

Table 8.1 Futures Orientations of Constitutions

Description	*Percentage of Constitutions*
No use of key terms	51%
One or two uses of key terms	31%
Several uses of key terms	10%
Numerous uses of key terms	6%
One explicit provision for advocating for future generations	2%
Two or more explicit provisions for advocating for future generations	0%

Approaches and suggestions for new national anticipatory institutions

Review of 100 constitutions did reveal, though, five approaches to amending them to create entities to advocate for future generations and also to judge whether nations are meeting their perpetual obligations. The first four approaches focus on amending current provisions in the constitutions. The fifth entails amending a constitution to create a new institution. Here are the five approaches:

(1) Amend provision for the commission or council on human rights to also advocate for future generations
(2) Amend provision for an ombudsperson to have responsibility to protect the rights of future generations
(3) Amend provisions for referenda, elected at-large representatives or nominated legislative representatives to represent the interests of future generations[6]
(4) Amend provisions to assign these responsibilities to extant specific institutions or persons
(5) Amend to create a new institution affiliated with the national court system

Each of these five approaches were considered for the 98 constitutions that did not contain an explicit provision. Before presenting the results of this exercise, please let me emphasize a couple of points. First, this assessment exercise is purely exploratory. Any proposal with respect to any country should be seen only as a seed of a proposal and the aggregated results are, similarly, meant to seed additional consideration of this issue. Second, I chose to identify the most straightforward path to constitutional amendments. For example, if there is a provision for a human rights commission, then I chose that option rather than to recommend the creation of a brand-new institution or giving the responsibility to a prominent person who is not mentioned in the constitution, though these are of course valid options.[7]

Table 8.2 presents the results of this exercise. Almost 40% of the constitutions have two very straightforward options available: (1) amendment of provisions regarding human rights commissions and (2) charging ombudsmen with this responsibility. Almost one-third of the constitutions offered an opportunity with a prominent existing institution or person. Almost one quarter of the constitutions had no provisional hooks and therefore would require the proposal of a new institution affiliated with the court system. Lastly, 8% of the constitutions encompassed an at-large representatives provision. Personally, I am surprised that over 75% of the constitutions have an existing provision that one could be straightforwardly built upon to establish responsibilities to future generations.

Table 8.2 Recommended Constitutional Amendments (by Type)

Recommendation	*Percentage*
Human rights commissions	21%
Ombudspersons	17%
Referendum or at-large representatives	8%
Other prominent institutions or persons specified in constitution	30%
New single-task institution as adjunct of court system	24%

Table 8.3 Recommended Constitutional Amendments by Country Descriptors (Means)

Recommendation	*Years Since Constitution Adopted*	*Years Since Constitution Amended*	*Population*	*Freedom Index*	*Human Development Index*
Referendum or at-large representatives	67	8.6	15,500,000	2.4	0.81
Human rights commissions	26	8.2	41,400,000	3.8	0.67
Ombudspersons	44	7.9	25,000,000	2.6	0.75
Other prominent institutions or persons specified in constitution	80	13.4	41,000,000	3.3	0.79
New single-task institution as adjunct of court system	58	13.2	70,000,000	2.9	0.77
Mean years	56	11.0	68,000,000	3.1	0.75
Median years	37	8	23,000,000	2.8	0.76
Minimum	2	2	38,000	1.0	0.40
Maximum	804	73	1,400,000,000	7.0	0.95

Table 8.3 presents some descriptive statistics related to these approaches. Two observations are highlighted. First, the institutions related to human rights and ombudsman are primarily associated with developing countries, which may be directly tied to a history of colonialism. This can be seen in the age of the constitution, Freedom Index, and Human Development Index columns. For example, countries with the newest constitutions are linked to the human rights commission approach.

A second observation relates to the years since a constitution was amended (second column). Here we see that constitutions have been recently amended (average of only 11 years, range 2 to 73). This observation suggests that constitutions can be viewed as living documents and that proposals to amend them with respect to future generations may not be doomed to the forces of inertia and conservatism.

Table 8.4 provides some examples of provisions in constitutions with respect to commissions or councils on human rights that could be responsible for future generations and making judgments about perpetual obligations. For example, the Kenyan Constitution formally establishes The Kenya National Human Rights and Equality Commission.[8] Its mission is "to protect and promote human rights through policy, law and practice" and to serve "as a watch-dog over the Government in the area of human rights." Consistent with the previous criteria, one option is to amend the Kenyan constitution to formally broaden the Commission's responsibilities to encompass future generations. This can be done by adding the function to the list of functions in Chapter 5 Section 59, subsection 2 of the constitution. Provision can also be made to designate a small number of Commissioners to represent the interests of future generations. Of course, the commissions would need to be revitalized and strengthened if they are found to be operating below envisioned expectations.

Table 8.5 presents some examples of countries whose constitutions include provisions for ombudsmen. Again, personally, I find it quite interesting that so many constitutions have provisions for ombudspersons.[9] Especially interesting to me is The Mediator/Ombudsman set out in the Moroccan constitution.

Table 8.6 presents examples of countries whose constitutions contain provisions for legislators who are not committed to representing specific, living constituents. For example, the Singaporean constitution contains a provision for nominated and non-constituency members of parliament.

Table 8.4 Examples of Commissions or Councils on Human Rights

Country	*Opportunity*
South Africa	South African Human Rights Commission
Nepal	National Human Rights Commission
Kenya	Kenya National Human Rights and Equality Commission
New Zealand	Human Rights Commission
Russian Federation	Commissioner for Human Rights

Table 8.5 Examples of Ombudspersons

Country	*Opportunity*
Morocco	The Mediator/Ombudsman
Papau New Guinea	Ombudsman Commission
Kosovo	Ombudsperson
Portugal	Ombudsman
Slovenia	Ombudsman for Human Rights and Fundamental Freedoms

Table 8.6 Examples of Referenda or At-Large Representatives for Future Generations

Country	*Opportunity*
Singapore	Nominate members of parliament to represent future generations
Ireland	Nominate members of the Senate to represent future generations
Norway	Elect at-large representatives to the Storting to represent future generations
Taiwan	Amend Article 26 to specify election of delegates to the National Assembly to represent future generations
Ukraine	Amend Article 72 to call for an All-Ukrainian referendum at a specified interval on whether current generations are meeting obligations to future generations
Afghanistan	Amend constitution to direct the President to appoint members to the House of Elders to advocate for future generations

Table 8.7 Examples of Assigning Responsibilities to Existing Prominent Persons or Institutions

Country	*Opportunity*
United Kingdom	House of Lords
Japan	Emperor
France	Defender of Rights
Myanmar	Constitutional Tribunal of the Union
Qatar	Prince's Advisory Council
China	Expand remit of The Chinese People's Political Consultative Conference
India	Encourage the Indian Supreme Court to periodically open a case to judge whether obligations to future generations are being met

Presumably, the constitution could be amended to specify that the constituency of the nominated members is future generations. The constitution of Norway allows the election of at-large representatives to its legislative body, the Storting. Possibly, this provision could be amended to also elect additional at-large representatives to represent the interests of future generations. In order to increase the influence of representatives of future generations, consideration could be given to providing each voter with more than one vote (i.e., one vote for their representative and one for an advocate for future generations[10]) or the power to form voting blocks to delay or block legislation deemed harmful to future generations.[11]

Table 8.7 presents some examples of assigning responsibilities to prominent persons or institutions. One intriguing possibility is to convince the House of Lords in the United Kingdom to accept the responsibility for assessing whether current generations of UK citizens are meeting their perpetual obligations.[12] A review by a Royal Commission of the role of the House of Lords in British society[13] found that the House of Lords should have these four main roles: (1) Provide a range of different perspectives; (2) Be representative of British society; (3) Provide checks and balances; (4) Engender perspectives. This mechanism for advocating for future generations meets the criteria established earlier, especially given the responsibility that the House of Lords accepts and oversees actions of the House of Commons and the government to protect the future well-being of the heirs of the United Kingdom. Subsequent to this proposal, another proposal has been put forth to establish an all-party parliamentary group on future generations within the House of Commons.[14]

The Indian constitution is over two-hundred pages long with text covering villages, social justice, and non-discrimination. There is no mention of future generations, foresight, etc. nor is there mention of councils or nominated parliamentarians. However, in India, one approach would be to amend the constitution to direct the Supreme Court to take up question of obligations to future generations periodically or even as needs emerge. This is a plausible proposal because the Indian Supreme Court can initiate legal proceedings on its own, instead of only having to deal with cases initiated by others. The downside of this approach is that the court already has 60,000 cases in the backlog.

Table 8.8 presents some suggestions for the establishment of new institutions. One idea I have explored for the United States is the establishment of The Court of Generations. The United States does not have a formal institution that has responsibilities for advocating on behalf of future generations. In fact, the only slight mention of future generations (i.e., Posterity) is found in the preamble of the U.S. Constitution. Given that mention of posterity in the preamble and the role of the Supreme Court in United States' governance, the proposal is to amend the U.S. Constitution to create a Court of Generations (CoG) that would reside in the judicial branch of government and would judge whether current generations are "*in contempt of intolerably threatening the security of the blessings of liberty to our Posterity.*"[15] The text of the proposed amendment is found in Exhibit 8.1.

Table 8.8 Examples of Creating New Institutions

Country	*Opportunity*
United States	Establish Court of Generations
Costa Rica	Establish a new autonomous institution to judge obligations to future generations
Brazil	National Congress establishes a new permanent committee to address future generation issues
South Korea	Assign responsibility to the Constitutional Court
Vietnam	Amend Article 102 to create people's court to judge obligations to future generations
Germany	Have Bundestag establish a committee of inquiry with respect to obligations to future generations

Germany modified the constitution, *Grundgesetz*, in 1994 to add environmental protection and included the term "generations": "Mindful also of its responsibility toward future generations, the state shall protect the natural basis of life by legislation and, in accordance with law and justice, by executive and judicial action, all within the framework of the constitutional order."[16] In 2006, a movement to add an article to the *Grundgesetz* that explicitly guaranteed intergenerational justice was unsuccessful. What is needed, and the approach that is consistent with the aforementioned criteria, is a new institutional 'representative of future generations' who could veto decisions or at least challenge them.[17]

Exhibit 8.1. Proposed Amendment to the U.S. Constitution to Establish Court of Generations

Section I. The power to judge threats to the security of the blessings of liberty to our posterity shall be vested in the Court of Generations, which shall be an adjunct of the judicial department of the national government.

Section II. The Court shall consist of a Grand Jury and the members of the Supreme Court. The Grand Jury shall return a bill of indictment to the members of the Supreme Court if evidence suggests an intolerable threat to the security of the blessings of liberty to our posterity. The members of the Supreme Court shall decide whether we and/or our ancestors are in contempt of intolerably threatening the security of the blessings of liberty to our posterity.

Section III. Each state and territory may appoint one person to sit on the Grand Jury.

Section IV. The Court shall convene within five years after the passage of this amendment and within every subsequent term of five years, unless specially convened by Congress, in such manner as Congress shall direct by law. Congress shall ensure that the Court has reasonable resources at its disposal to assist its deliberations.

World Court of Generations

It is proposed that a World Court of Generations (WCG) be established to periodically judge whether humanity is meetings its perpetual obligations and if not, what actions are needed. Ideally, this judgment would be made with the benefit of inputs from NAIs from around the world. I can imagine that the WCG would institute a process to engage NAIs in making this judgment, especially when there are sharp disagreements across NAIs about what perpetual obligations are not being met and which countries might be more or less responsible for the shortcoming. Practically, in the short term, the WCG will be responsible for making this judgment for humanity on its own until a substantial number of NAIs are up and running.

There are many issues to address with respect to the design and establishment of the WCG. One can imagine the responsibility for forming the WCG would fall to the United Nations, maybe UNESCO given its commitment to future generations. Similar to the IPCC, the WCG would periodically report its judgments to humanity. The exact nature, composition, and processes of the WCG would subject to negotiation with UN member states. The budget for the WCG would also be a point of negotiation amongst member states.

The WCG would operate as a court in the broadest sense of the word. The WCG would allow the presentation of information for the official and historical record. The WCG would allow arguments about what this information means with respect to its assessments about how well, if at all, each of the perpetual obligations is being met. These arguments could challenge the evidence as well as the assessments. So, one can imagine that rigorous debates could focus on the evidence and actions pertaining to avoid human extinction, especially if the evidence suggests that Manhattan-scale projects or more pervasive actions are needed. All of the proceedings would be streamed live around the globe with translations in as many languages as possible.

Intergenerational panel on perpetual obligations

A boundary organization is a neutral party that facilitates meetings of stakeholders and provides technical expertise and information to stakeholders.[18] In this case, a boundary organization is needed to provide information to the global base of stakeholders (i.e., GADSS users, the organizations discussed in this section) on the metrics that define our perpetual obligations to future generations. Institutionally, one can imagine a boundary organization whose focus is on perpetual obligations to future generations as analogous to the Intergovernmental Panel on Climate Change (IPCC). The IPCC is the boundary organization that is the neutral body charged with collecting all pertinent evidence and distilling and synthesizing the data for use by policy makers and citizens worldwide. IPCC is a transformative organization because of these responsibilities and because it is unique in taking a long-term view in its analyses, out to the year 2100 and beyond.[19] Other global boundary organizations include the World Health Organization, the International Energy Agency, and the Food and Agricultural Organization of the United Nations.

Let's call the body charged with assembling the data to allow the NAIs and WCG to assess whether perpetual obligations to future generations are being met the InterGenerational Panel on Perpetual Obligations (IPPO) (see Figure 8.1). The IPPO would operate similarly to the IPCC in that it

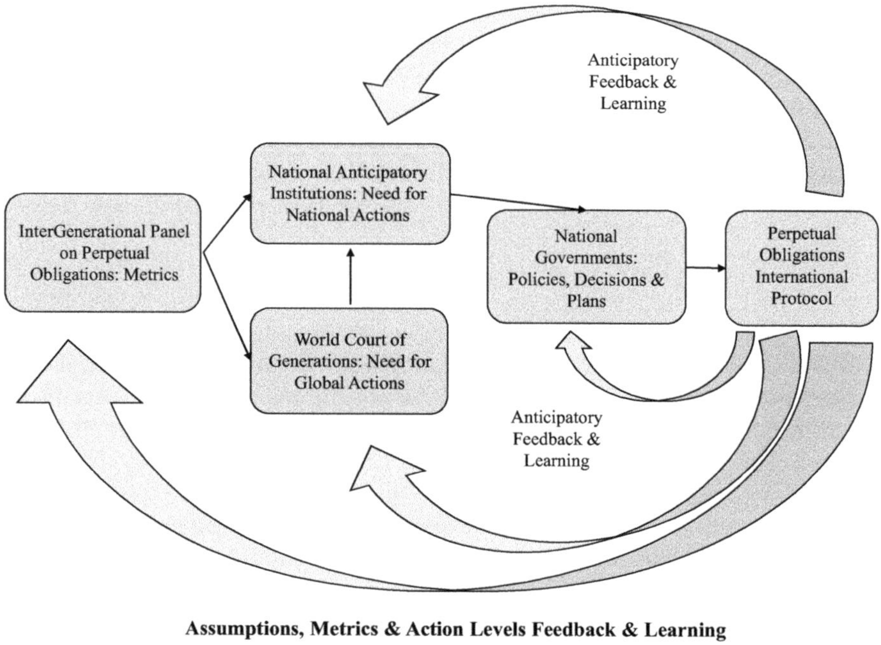

Figure 8.1 Global and National Decision-Making About Meeting Perpetual Obligations

would be responsible for assessing humanity's knowledge in the many and diverse areas encompassed by all dozen obligations. The IPPO would need to work hard to explicitly define rigorous and credible metrics related to each of the obligations.

The IPPO will have several responsibilities. Among them would be the responsibility for facilitating the development, administration, and oversight of the Whole Earth Monitoring System (WEMS) introduced in Chapter 4. The IPPO would also have similar responsibilities for the GADSS and ensuring that the use of these systems is seamless for users worldwide.

The IPPO will also be the keeper of the themed perpetual obligation scenarios.[20] It will administer the crowdsourced effort to develop the scenarios. It will vet the scenarios for internal validity and plausibility. It will make the scenarios available to the world. It will develop tools to assist scenario developers and users of the scenario database. The IPPO will do this without usurping or stunting the development of a myriad of other scenarios, such as those developed by the Millennium Project,[21] International Energy Agency,[22] and the IPCC itself.[23] These additional scenarios serve important purposes and can even find their use in establishing perpetual obligation metrics.

To complement and even improve scenario development and writing, the IPPO would be responsible for scanning for disruptive technologies that could further bear upon the obligations related to the essence of nature and humanity. The IPPO would also need to develop signposts designed to provide early warning signals with respect to each of the obligations.

With respect to Figure 8.1, the IPPO would be responsible for using data and information from both systems and the scenario repositories to estimate metrics related to each of the twelve perpetual obligations. For example, the IPPO would be responsible for estimating the probability of human extinction (e.g., by using Singular Chain of Events Scenarios) and using evidential reasoning approaches to assess the more qualitative metrics that are more appropriate for other perpetual obligations (e.g., with respect to preserving human knowledge). The IPPO would be an active participant in international dialogues about meeting our perpetual obligations as it would need to receive, aggregate, and process stakeholders' anticipatory decisions to assess to what degree perpetual obligations might then be met given the intentions regarding national plans, programs, and policies.

Lastly, the IPPO would also be responsible for the foresight and communications aspects related to Perpetual Obligation #12: fully inform current generations about futures and obligations. Its primary responsibility here would be to effectively communicate assessments with respect to meeting perpetual obligations. The IPPO would be responsible for developing the communication and public participation tools.

National governments and meeting perpetual obligations

Provisions for NAIs are embedded within national constitutions, yet these institutions are not given responsibilities normally allocated to the three main branches of government. They cannot pass laws or authorize expenditures. They cannot execute the wishes of the legislative branch nor enforce the laws or adjudicate cases, with few exceptions such as the proposal for the Supreme Court of India mentioned earlier. This design principle is to keep the NAIs out of everyday politics so that as much as possible their messages are not tainted by partisan bias. Optimistically, the normal channels of government could fund programs and policies to meet perpetual obligations. Conversely, allocating programmatic and expenditure powers amongst two separate components of government could have disastrous unintended consequences for both current and future generations.

I admit that relying upon regular governments to approve and administer policies and programs to meet perpetual obligations is a weakness in this scheme. In the United States, changes in political administrations result in substantial changes in funding for environmental protection and administration and prosecution of environmental laws.[24] Such changes would not provide the consistency needed to meet perpetual obligations of any kind and could be viewed as flaws in systems of governance. The next chapter addresses this situation by proposing that humanity take on a very long program of experimenting with different forms of governance systems that could represent improvements over today's systems.

The perpetual obligations international protocol

Decisions made by national governments (and subgovernments) across the globe will largely determine whether humanity will meet its perpetual obligations. It is recommended that these decisions be formalized in the Perpetual Obligations International Protocol (POIP). The POIP would be similar to the Montreal Protocol on Substances that Deplete the Ozone Layer[25] and the Kyoto Protocol.[26] The POIP would have twelve parts, each part relating to one of the twelve obligations. Each part would set out each country's commitments toward satisfying that particular perpetual obligation. The WCG would be responsible for facilitating and mediating discussions leading toward agreements. The WCG would make full use of the GADSS and inputs from the IPPO in this role. The United Nations could also establish an Intergovernmental Negotiating Committee to assist the WCG in crafting the POIP.[27]

The WCG would not be charged with determining whether commitments across the board for meeting each perpetual obligation are fair or not. The WCG will only be concerned about the end product of the discussions. Issues of fairness will certainly infuse the multiparty discussions about each perpetual obligation and the ultimate result of these discussions will reveal what fairness means at that particular time and place.

Once finalized, its terms would be binding, although appeals could be made to the WCG if terms are thought to be overburdening, biased, or unjust. The POIP would be considered perpetual: though it will need to be revisited and updated relatively frequently, it will never expire. In other words, responsibilities for meeting obligations to future generations will always be present even as the composition of the global community of nations changes.

No one should be under the illusion that negotiating the POIP would be straightforward or easy in any sense. Certainly, international dialogue and agreements to take action to mitigate climate change have been extraordinarily difficult and have not met expectations. With respect to meeting perpetual obligations to future generations, determinations with respect to culpability could be much more complex (e.g., risks of human extinction may include climate change and also pandemics, nuclear weapons, asteroids, etc.).

Agreements to deal with this obligation, therefore, may be composed of numerous interrelated sub-agreements. One could ask that as a matter of course with respect to the POIP whether it is ever justifiable for the international community to forcibly intervene in countries that are especially culpable in situations where obligations are not being met. After all, the world community has intervened in the affairs of other countries for humanitarian reasons, such as in Kosovo, Somalia, and Haiti. I explored the question of intervention with respect to unsustainable energy policies. As could be expected, it is found that successful active interventions are generally implausible, infeasible, and inadvisable if those countries most egregiously in violation of accepted principles are large and have potent militaries.[28] Thus, it is not recommended that the WCG or any other body have the police powers to enforce the POIP.

Lastly, the IOIP would cover numerous topics. It could pave the way for additional new intergovernmental organizations that may have very specific responsibilities. For example, it could help establish an institution capable of being the steward of the world's nuclear wastes through economic and political instabilities, which can be referred to as perpetual organizations (see the following section). It could also encompass numerous existing national and international specialty organizations, such as the European Space Agency and the U.S. National Space and Aeronautics Administration with respect to grand challenge projects involving space and maybe even a large number of private sector firms, such as Google and Amazon with respect to archiving humanity's knowledge base.

Perpetual organizations

Lastly, perpetual organizations are needed to ensure that critical responsibilities to humanity are met without interruption. It is optimistic to assume that the institutions introduced earlier will be able to weather severe global political, economic, and cultural upheavals. Such upheavals may not lead to our extinction, but they could shut down most aspects of civilization and re-prioritize efforts for extended periods of time. In anticipation of this possibility, humanity needs to establish a small number of organizations that are designed to survive upheavals. Indeed, they will need to be materially self-sufficient so that their existence is not dependent upon a world in flux. Three perpetual organizations are proposed here.

A close look at history reveals numerous human institutions that have survived for hundreds of years and even thousands of years. Many of these institutions are religious, such as the Catholic Church. Examples of current monastic orders founded many years ago, include: the Dominican monastic order or Order of Friars Preachers that was founded in 1216 through the leadership of Santo Domingo De Guzman or Saint Dominic; the Franciscan monastic order that was founded in 1210 by Francis of Assisi, which called for a contemplative life combined with the active duties associated with secular clergy; the Jesuit order, or Society of Jesus, that was founded in 1540 under the leadership of Ignatius of Loyola; and the Sangha community of Buddhist monks that was founded by Siddhartha Guatama around 500 BCE.[29] A less well-known group is the Falashas Jews, who are descendants of Canaan Jews who migrated to Ethiopia and maintained their religion despite isolation from the larger worldwide Jewish community for over 2000 years and in spite of persecution from others in Ethiopia.[30]

These and other long-lasting religious institutions share many characteristics. First, their mission is clear: they exist to maintain a religious community. In a sense, they are stewards of the faith in their communities. Second, members of these institutions believe that the missions are worthy of devotion and sometimes huge personal sacrifices. Third, they have evolved some manner to support themselves. The Christian monastic institutions mentioned earlier are largely self-sufficient. On the other hand, the Sangha depend wholly upon the generosity of the laity, a generosity that is well ingrained into the social fabric of the larger community. Many religious institutions are stewards of places and relics – for example, the Western Wall, the Dome of the Rock, and the Shroud of Turin. However, none hold stewardship responsibilities analogous to those discussed herein.

There are several other examples of long-lasting institutions. Many universities have admirable histories, including the University of Paris, founded about 1170, the colleges established at Oxford, including University College (1249), Balliol College (1263), and Merton College (1264), and also the Jagiellonian University in Cracow, Poland, founded in 1364. Institutions embodied in the establishment of law also have long histories. For example, maritime law, while possibly first being developed by the Egyptians or Phoenicians, can be more definitely attributed to Rhodes about 2000 years ago, as testified to by surviving written comments from Roman emperors. Human institutions associated with indigenous cultures can sometimes be traced by very long periods of time. For example, the N/um chai is a curing ceremony trance dance practiced by the Bushman of the Kalahari that can be traced back approximately 40,000 years [16].[31]

In summary, there is a precedent for long-lasting organizations already. Let's assess suggestions for three perpetual organizations.

Stewardship Institution[32] – The Stewardship Institution (SI) would be given the responsibility for overseeing the world's nuclear wastes. As noted in Chapter 2, these wastes will be harmful to earth life for approximately 10,000 years. Such a long time frame can easily conjure visions of depopulated landscapes and abandoned human settlements that are the purview of archaeologists worldwide. The essential element of this vision is that nuclear waste repositories could be forgotten, and future humans could stumble into them and be at risk. To deal with this scenario, researchers have developed symbology that, it is hoped, could communicate the dangers across millennia.[33]

Given that there are over seven billion humans alive on the planet and that virtually every area of the planet is inhabited, it is hard for me to imagine the depopulated landscapes and abandoned human settlements scenario on such a wide scale that areas of nuclear waste repositories would be forgotten. Instead, I can imagine upheaval leading to situations where institutions managing the repositories collapse, along with the knowledge about the repositories. The stewards in the SI would live totally self-sufficient lifestyles while also maintaining the sites, knowledge about the sites, and protecting humans and other earth life from the dangers of the sites.

Order of Perpetual Obligation #9 Monks – The Knowledge Monks would be entrusted with humanity's knowledge. Whilst they may not be equipped with the quill and parchment given to monks from the Middle Ages, they will essentially have the same task. They will also need to be self-sufficient lest upheaval results in the destruction of the world's knowledge repositories. Establishment of the Knowledge Monks is just as essential to meeting Perpetual Obligation # 9 as is the technology needed to preserve the knowledge.

Farsighters – The Farsighters would continuously manage humanity's exploration for other homes in our galaxy. Specifically, Farsighters would continuously monitor data returning to the earth from probes dispatched to explore planets in Goldilocks zones around suns across our galaxy. Since it could literally take centuries or longer for probes to reach their destinations, humanity ought to have a dedicated group of individuals who most certainty be able to weather upheavals to ensure that the probes are not forgotten and that their reports are properly logged and maintained for humanity's use.

Notes

1 D. Thompson. 2010. Representing Future Generations: Political Presentism and Democratic Trusteeship. *Critical Review of International Social and Political Philosophy*, 13, 1, 17–37.

2 Institute for European Environmental Policy. 2015. Establishing an EU 'Guardian for Future Generations'. Report and Recommendations for the World Future Council.

3 M. Göpel and M. Arhelger. 2010. How to Protect Future Generations' Rights in European Governance. *Intergenerational Justice Review*, 10, 1, 4–10.

4 www.constituteproject.org

5 "Sustainable development is development that meets the needs the present without compromising the ability of future generations to meet their own needs." Brundtland Commission. 1987. *Our Common Future*. Oxford University Press, Oxford.

6 This is sometimes referred to as a restricted franchise model. A. Dobson. 1996. Representative Democracy and the Environment. In W. Lafferty and J. Meadowcroft (Eds.), *Democracy and the Environment*. Elgar, Cheltenham, 124–139.

7 It should be noted that just because a provision exists in a constitution does not mean that it is currently being implemented even without a future generations perspective.

8 www.knchr.org/About-Us/Establishment

9 For a review of ombudsperson provisions, see http://comparativeconstitutionsproject.org/files/cm_archives/ombuds.pdf?6c8912

10 K. Ekeli. 2005. Giving a Voice to Posterity – Deliberative Democracy and Representation of Future People. *Journal of Agricultural and Environmental Ethics*, 18, 429–450.

11 K. Ekeli. 2009. Constitutional Experiments: Representing Future Generations Through Submajority Rules. *The Journal of Political Philosophy*, 17, 4, 440–461.
12 This idea was first presented in this article: B. E. Tonn and M. Hogan. 2006. The House of Lords: Guardians of Future Generations. *Futures*, 38, 115–119.
13 Royal Commission on the Reform of the House of Lords. 2000. A House for the Future, January. Retrieved from www.archive.official-documents.co.uk/document/cm45/4534/4534.htm.
14 N. Jones, M. O'Brien, and T. Ryan. 2018. Representation of Future Generations in United Kingdom Policy-Making. *Futures*, 102, 153–163.
15 B. E. Tonn. 1991. The Court of Generations: A Proposed Amendment to the U.S. Constitution. *Futures*, 23, 5, 482–498.
16 The Federal Republic of Germany. 2019. Basic Law. Retrieved from Germany, The Federal Government, www.bundesregierung.de/breg-en/chancellor/basic-law-470510
17 M. Schröder. 2011. The Concept of Intergenerational Justice in German Constitutional Law. *Ritsumeikan Law Review*, 22, 321–330.
18 D. Guston. 2001. Boundary Organizations in Environmental Policy and Science: An Introduction. *Science, Technology, & Human Values*, 26, 4, 399–408.
19 B. E. Tonn. 2007. The Intergovernmental Panel for Climate Change: A Global Scale Transformative Initiative. *Futures*, 39, 614–618.
20 See Exhibit 6.2 for a list of the themed scenarios.
21 www.millennium-project.org
22 World Energy Scenarios. Retrieved from www.iea.org/reports/world-energy-model
23 For example, its mitigation pathways, www.ipcc.ch/site/assets/uploads/sites/2/2019/02/SR15_Chapter2_Low_Res.pdf and emissions scenarios, www.ipcc.ch/report/emissions-scenarios/
24 N. Vig and M. Kraft. 2013. *Environmental Policy: New Directions for the 21st Century*. CQ Press, Thousand Oaks, CA.
25 www.unenvironment.org/ozonaction/who-we-are/about-montreal-protocol
26 https://unfccc.int/kyoto_protocol
27 See for instance: D. Feldman. 1994. Iterative Functionalism and Climate Management Organizations: From 'Intergovernmental Panel on Climate Change' to 'Intergovernmental Negotiating Committee'. In R. V. Bartlett, P. A. Kurian, and M. Malik (Eds.), *International Organizations and Environmental Policy*. Westport, CT: Greenwood Press, 189–209.
28 B. E. Tonn. 2011. Intervention in Countries with Unsustainable Energy Policies: Is It Ever Justifiable? *Futures*, 43, 3, 348–355.
29 R. Lester. 1993. Buddhism: The Path to Nirvana. In H. Byron Earhart (Ed.), *Religious Traditions of the World*. Harper Collins, San Francisco, CA, 847–972.
30 D. Kessler. 1996. *The Falashas: A Short History of the Ethiopia Jews*. 3rd ed. Frank Cass, London.
31 J. Tassel. 2000. Yo-Yo Ma's Journeys: Making Music with Humanity, From Sanders to the Silk Foad. *Harvard Magazine*, 102, 4, 42–51.
32 B. E. Tonn. 2001. Institutional Designs for Long-term Stewardship of Nuclear and Hazardous Waste Sites. *Technological Forecasting and Social Change*, 68, 255–273.
33 G. Benford. 1999. *Deep Time: How Humanity Communicates across Millennia*. Avon Book, New York; S. Hora and D. Von Winterfeldt. 1997. Nuclear Waste and Future Societies: A Look into the Deep Future. *Technological Forecasting and Social Change*, 56, 2, 155–177; J. Lomberg and S. Hora. 1997. Very Long-term Communication Intelligence: The Case of Markers for Nuclear Waste Sites. *Technological Forecasting and Social Change*, 56, 2, 171–188.

9 Passing through a socio-cultural singularity

Futurist Ray Kurzweil envisions that humanity will soon enter and pass through a Singularity caused by the synthesis of exponentially changing information, bio-, and nano-technologies.[1] The metaphor is based on an unknowable characteristic of a black hole; if one were to enter a black hole what would wait for the intrepid traveler on the other side? This is an apt metaphor for this chapter as well. I believe that humanity must traverse a socio-cultural singularity in the next 1000 years to avoid cascading global catastrophes. Failure to prepare for this could result in humanity becoming another data point supporting the verisimilitude of the Drake Equation.

Yes, technologies will play an important role in the Singularity but not the only role and not even the most essential. This is because fundamental changes are needed in the most fundamental building blocks of society: systems of governance; economics; religion; culture; and psychology.[2] These changes are needed to assuage fears people may have about their own lives and livelihoods as humanity and its institutions evolve toward being more futures-oriented. These changes will also directly support meeting perpetual obligations. For example, the discussion on systems of governance directly relates to Perpetual Obligation #3, Bequeath Sustainable Societies and the one on economics directly relates to Perpetual Obligation #4, Bequeath Sustainable Production.

This chapter should be viewed as a very small step toward the Singularity. It presents critiques of society in each of these five areas and then also numerous seeds for thought and indicative changes that address the critiques. It is suggested that a set of Transformative Scenarios be developed to help humanity better anticipate and shape the coming Singularity. The chapter concludes with an example of a Transformative Scenario.

Governance

The institutional innovations presented in Chapter 8 are necessary but not sufficient to create sustainable societies that are capable of meeting perpetual obligations and making effective and timely decisions to ensure humanity's journey. This is because the institutions introduced in Chapter 8 are designed to advocate for future generations and act as our conscience when we are not meeting our perpetual obligations. These institutions do not have the power to fund and implement policies and programs. These powers are assumed to still be resident with our everyday political systems. And, given the scorecards and other assessments thus far presented, the current institutions have not been up to the job. It is an open question as to whether current institutions fail primarily because of their design or primarily because the people who are ensconced in them are defective, selfish, and prone to myopia. We do know that it is easy to mis-attribute blame to individuals when in fact situations can make good people behave badly (see the famous Milgram electroshock[3] and Stanford prison experiments[4]). In this section, let's assume that the institutions themselves share a good deal of blame.

The first question, then, is whether current political institutions can evolve to be more futures-oriented, to be able to meet the spirit of the third perpetual obligation introduced in Chapter 6, Bequeath Sustainable Societies. If the answer is no, then what new types of political institutions ought humanity turn to?

A hopeful answer to the first question is yes. The development of national anticipatory institutions introduced in Chapter 8, combined with reasons to care about future generations, perpetual obligations, and their metrics as guides, will be enough to foster increased futures-orientation of existing governance institutions. Moreover, existing systems of government were primarily designed to referee the lives of living human beings. Most constitutions fail to address the future at all. Therefore, the more realistic answer to this question while evolution may be possible, a revolution in the design of systems of government may be needed.

In other words, it can be argued that our systems of everyday governance may need to pass through the Singularity. We have not yet arrived at the end of history, as Fukuyama famously declared when capitalism seemed to win out over communism when the Soviet Union fell.[5] More specifically, democracy as practiced today is probably not the end of political history. We cannot fossilize our current myopic political systems and survive. In fact, over the course of the next thousand years or so, new and more effective political systems will need to be designed, tested, implemented, and evaluated. We cannot learn what might be better without political system experimentation. This sounds scary, and it is, as this is our first peek into the socio-cultural Singularity.

To successfully traverse the singularity, much will need to be learned over time about potential new political systems. Humanity should collectively decide to implement innovative and long-term experiments to gently but effectively test new socio-political-economic institutions, instead of engaging in wars (figuratively and literally) about whose systems of what type are best.[6] These experiments could take many years to implement and many more years to evaluate. Theoretically, we are experimenting right now with only a limited number of models of governance, none of which seem to be working well from a futures perspective. Fortunately, humanity has time for more robust, systemic, and thoughtful experimentation leading up to and past the Singularity, if we don't destroy ourselves in the meantime. We especially need to experiment with ways of fostering political change that does not result in domestic violence or international wars. Old states need to be allowed to sunset, dissolve, and reemerge in different forms and combinations of populations without recourse to death and destruction that could threaten the journey or lead to our own extinction.

Here are a couple of new models of governance to consider to stimulate creative thought and serious discussion of better ideas. The first idea is labeled non-spatial governments (NSG) and the second is referred to as a futures congress.

Non-spatial governments are governments that share geographic space but do not have defined spatial boundaries.[7] These governments could be useful when people who identify with each other culturally are separated by artificially demarcated geographical jurisdictional boundaries and also when these people are minorities and therefore are under risk or worse from violent ethnic cleansing. Each NSG would have its own political institutions, laws, policies, and social programs. The NSGs in a region would share responsibilities for major infrastructures (e.g., electrical, road, water systems) and cooperate to enforce laws in overlapping situations. Current information and communication technologies can support political participation in NSGs and the delivery of services, in addition to the Global Anticipation and Decision Support System described in Chapter 7. How these technologies can support a non-spatially defined court system[8] and much more sophisticated and personally tailored political participation[9] have also been explored.

The second idea is referred to as *futures congress.*[10] This political institution was designed to meet these nine criteria:

(1) Explicit recognition of future generations and future-oriented issues;
(2) Explicit implementation of a structured decision-making process;
(3) Bias toward consensual decision-making;
(4) Incentives to include people of wisdom;
(5) Effective and broad-based citizen participation;
(6) Limitation of special-interest lobbying;
(7) Ability to balance long-term and short-term interests;
(8) Ability to make stable commitments to long-term plans and actions;
(9) Ability to foster learning.

Table 9.1 presents the framework for a futures congress using the approximate population of the United States as a model.[11] The process starts with citizens choosing a Decision Maker to vote on legislation on their behalf. Each Decision Maker needs to have support of between 100 and 200 citizens.

Table 9.1 Chambers of the Futures Congress

Chamber	*Duties*	*Requirements*	*Decision Process*	*Number*
Elders	Sets out criteria to guide the creation and evaluation of future-oriented decisions.	≥ 60 years of age ≥ 5 years as a Visionary Commitments from 10 to 20 Visionaries	Consensus	~20
Visionaries	Creates future-oriented decision alternatives. Each Visionary uses moral and ethical judgment to commit to an Elder.	≥ 50 years of age ≥ 5 years as a Realist Commitments from 100 to 200 Realists	Brainstorming, Consensus	~200
Realists	Evaluates future-oriented decision alternatives. Each Realist uses moral ethical judgment to commit to a Visionary.	≥ 40 years of age ≥ 5 years as a Decision Maker Commitments from 100 to 200 Decision Makers	Delphi	~20,000
Decision Makers	Choose among future-oriented decision alternatives. Each Decision Maker uses moral and ethical judgment to commit to a Realist.	≥ 30 years of age Commitments from 100 to 200 citizens	Approval voting	~2 million
Citizenry	Each citizen uses moral and ethical judgment to commit to a Decision Maker.	Legal citizenship ≥ 18 years of age	Individual commitment	~200 million

This number, sometimes called the Dunbar number,[12] allows each Decision Maker to actually know their supporters and for the supporters to know each other. Similarly, Realists need support of about 200 Decision Makers, Visionaries the support of about 200 Realists, and Elders the support of about 10–20 Visionaries to do their jobs of assessing policy alternatives, creating policy alternatives, and guiding long-term policymaking, respectively. Within the context of this book, the Elders would in turn be guided on the journey by perpetual obligations and findings from the national anticipatory institutions.

In summary, one can argue that major transformations of systems of governance could be needed to traverse the upcoming Singularity. Not only must day-to-day systems of governance become more futures-oriented but meta-systems are needed to foster non-violent changes of political systems. This is exceedingly important if change arises from grassroots movements that challenge those with entrenched political power. The prospects that disruptive political change could result in civil, regional, and world wars could thwart the evolution of new systems. In addition to great loss of live, violence could endanger programs to meet our perpetual obligations, though humanity can also establish a limited number of perpetual organizations that could weather such disruptions.

Humanity ought to consider experimentation with new systems of governance. Evaluation of governments over time will be intellectually and ethically challenging. How long should an experiment persist until one can judge a new form has failed or not yet failed? A century, two centuries, a millennium? What does failure mean? One can assume that if a nation is failing to meet its obligations to its present generations,[13] it will also be failing to meet its perpetual obligations. However, within the big picture, a state will fail if it also fails to meet its perpetual obligations to humanity. Presumably, it will be possible to track nations' commitments to meeting each perpetual obligation over time. A pattern of insufficient commitments would indicate a failed state, though much thought and discussion are needed to determine the thresholds to support this judgment. Maybe ideas developed in Chapter 7 with respect to metrics, upper and lower probabilities, and evidential reasoning could also be used for this task.

Economics

Modern economics and resulting economies are a major barrier to futures-oriented thinking. This is not a new revelation for many who work to protect the environment, improve the sustainability of production, and care about meeting obligations to future generations. The complaints generally fall into these categories:

- The incessant goal for economic growth is not environmentally sustainable and even the notion of economic growth over the next 1000 years makes little sense (i.e., holding population steady, what does a 3% growth rate mean after a thousand years, that each individual is consuming products and services at light speed?).

- The way economic growth is measured (i.e., gross domestic product) rewards policies that favor consumption over policies that may moderate consumption to stave off extinction-level risks such as climate change.
- Monetary policy is designed to foster full employment while also controlling inflation. In times of economic downturns, central banks adjust monetary policies to increase economic activity, primarily through the private sector and primarily through increased consumption. It is surely a bias to only focus on the private sector given the substantial role that the non-profit sector plays in modern economies. Consumption for consumption's sake works against policies designed around sustainability and environmental goals.
- The dominant model used to make economic decisions, benefit–cost models, discount future benefits and focus on near-term goals. Even with a modest discount rate of, say 3%, which some argue should be the social discount rate,[14] the weight of future benefits of any social investments converge to zero within very short-term horizons.
- All economic decision-making, basically, is transacted in monetary units, such as dollars. Thus, economists work hard to find ways to put monetary values on items, such as clean water and indigenous ways of life, simply to be able to compare investment costs with monetized benefits. Contingent valuation approaches are controversial[15] as are approaches to monetize ecosystem services.[16] It seems quite impossible to render very long-term decisions about perpetual obligations using only monetary units. This is one reason why obligatory levels of actions addressed in Chapter 7 are based on qualitative and quantitative assessments that are not monetary in nature.
- Micro-economic models are based on the notion of individual preference or utility functions. Policy makers are admonished to not make decisions without knowing what consumers really want. Since it is impossible to know with certainty the preference functions of future generations, many economists tend to dismiss policy making that could benefit future generations. We have dealt with this issue in Chapter 5, but it bears repeating: this is an untenable and fatal flaw in economic theory as applied to futures decision-making.
- Arguments to save jobs at the expense of irreplaceable environmental amenities, human health, and climate catastrophe present a false choice. Creative monetary and fiscal policy can create jobs quite expeditiously and can create the training programs to aid people adapt to new markets. Jobs per se are never lost as they are easily replaceable. The near term does not have to dominate longer-term concerns in this area.

- Why is it that market societies cannot pay for public goods without the revenues generated by taxing the private sector, sales of goods and services, and personal income? Why must funding for schools be cut when consumption of, say, airline services or gasoline goes down, when the human and physical infrastructures are still intact? The need to disentangle consumerism from the providence of public services is especially acute when addressing the meeting perpetual obligations.[17]
- Markets that are dominated by computer trading and operate at sub-second speeds have absolutely no relationships to futures-oriented policy making.
- As economists freely admit, and then ignore, their models do not incorporate equity considerations. This is a fatal flaw in the context of the journey and obligations to future generations.
- Lastly, economic models that place the individual consumer at the heart of reality promote a single-minded individualism that is at odds with the duty to help meet perpetual obligations and feelings of solidarity with all generations. It promotes a self-defeating selfishness that, I believe, works against both individual psychological health and also against the program set out in this book.

In summary, the major pillars of modern economics seem to be at odds with the futures program laid out in this book. Dominant economic theory and practice with respect to policy are too short term to completely serve humanity's very long-term interests. This is especially a problem when economic methods and biases are used to make decisions that ought to be long term. As has been pointed out by others, economies exist within societies, which exist within the earth's global environmental systems, as shown in Figure 9.1,[18] though decision-making often has this reality reversed. Thus, economic models and methods currently used to make decisions which have long-term consequences need to be supplanted by the set of ideas presented in Chapters 5–8.

This does not mean that economic theory, models, and methods ought to be jettisoned entirely. There will still be decisions regarding the allocation of scarce resources and maintenance of meaningful employment that provides for a robust quality of life. Markets can express our values for empowerment and entrepreneurism. Economics provides numerous important ideas, including, and not limited, to: comparative advantage; sunk and opportunity costs[19]; marginal costs; monopolies and monopsonies; auction theory; and behavioral economics (or policy nudges discussed in Chapter 7). And as mentioned earlier, societies now and in the future could choose to have market economies. Internal rates of return could still drive day-to-day economic decisions. But, the demands of meeting perpetual obligations will constrain modes of production, at least in the short term. As expressed in previous chapters, humanity has not yet successfully

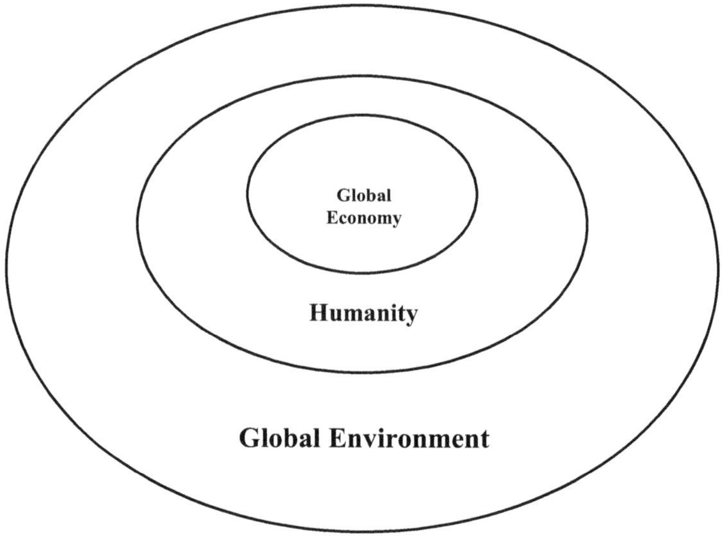

Figure 9.1 Conceptual Diagram of Sustainability

constrained the use of non-renewable resources or reduced the emissions of GHGs to prudent levels.

It is also anticipatable that economics will be reimagined on the other side of the Singularity. For example, one way to delink monetary policy from consumerism would be to allow central banks to temporarily increase resources available to philanthropic foundations and charities to expend directly in their communities. One mechanism could be to deposit funds in accounts held by those organizations at central bank institutions. Such a policy would complement fiscal policies, such as block grants, designed to address economic downturns that mainly impact middle-class and economically disadvantaged households and communities.

Another peek into the Singularity with respect to economics relates to a major disruption in production. Advanced distributed technologies that are currently disrupting several traditional economic sectors may synergize with extremely affordable technologies being developed for bottom-of-the-pyramid countries to create affordable and effective household- and community-based self-sufficient systems of production.[20] Envision households that would be relegated to being economically disadvantaged in brutal market economies being able to self-produce twice what they could have purchased with 50% of their annual household income.

Self-sufficient systems would have many components, including: distributed energy technologies such as rooftop photovoltaics and solar water heating; 3D printing that could use recycled materials to produce just about any household item, from utensils to clothes; extensive 3D printing systems that could print homes and even automobiles;[21] local agriculture to produce food, fuel, and agriceuticals;[22] and water purification and recycling systems. The One-Laptop-Per-Child initiative demonstrated that it is possible to manufacture inexpensive, functional, and rugged computers that use open-source software and mesh computing to link to the internet. One could imagine that these systems could become indefinitely sustainable through renewable, reusable, and recyclable practices.

What I like about this vision is that it still emphasizes entrepreneurship, as well as empowerment, creativity, sustainability, and new markets that emerge very locally. As I am writing this section, the world is in the grips of the Coronavirus crisis. In the United States where I live, the economy is basically shutting down. Tens of thousands of service workers and others face losing their jobs and income. Their homes and neighborhoods are virtually un-self-sufficient. Imagine the resiliency to the U.S. economy had homes and communities across the country been equipped with self-sufficiency systems described earlier.

This vision of an economy on the other side of the Singularity disrupts numerous conventional ideas and notions of economic theory and public policy. Here are a few:

- What does unemployment mean when households, at least theoretically, can meet 50% or more of their economic needs through self-sufficiency systems? How can 'unemployment' be measured? What macroeconomic policy levers might be developed and deployed to ensure full employment? What does economic growth mean in this new economy, if anything? Or are we now to focus on economic development, and if so, how should that be measured? What measure would be needed to replace GDP?
- How should governments cope with a reduction in tax revenue from sales and income taxes? In the United States, how would Medicaid and Medicare be fully paid for?
- What does social security mean in this world? How is it paid for? Maybe recipients are required not only to contribute financially but also through the accumulation of outputs from self-sufficient systems over their lifetimes, which then can be drawn down after their 'retirement' age?
- Lastly, does capital take on a different meaning in a highly self-sufficient world? What does human capital mean? What does wealth mean?

I am sure that there are numerous additional economic considerations with this new model of the economy. It seems to me that the revolution in economic theory could rival the shift from Newtonian mechanics to quantum mechanics and the general theory of relativity in physics.

Religion

Similar to economics, there are several religious tenets that are fundamentally incompatible with the program set out in this book. The set includes:

- *Determinism* – This belief holds that all paths into the future are predetermined by a deity.[23] Devotees place their trust in the deity. If things work out, fine. If things do not work out, it was at the deity's will. This belief does not value anticipation and proactive behavior. It is a disempowering belief, as it also releases devotees from worrying about the long-term stewardship of the earth and relieves devotees from any concerns they may have about their obligations to future generations, maybe with the exceptions of their own children and grandchildren.
- *Fatalism* – This belief is different from determinism but has essentially the same result. Fatalists, whom by-the-way do not have to find their fatalism in religious dogma, believe proactive actions in their own self-interests and for the greater good are bound to fail. The stars never align. Anticipation is a waste of time as it will never produce useful and actionable insights. Efforts to sustain life on earth will not stave off extinction. Even considering obligations to future generations is an act of futility.
- *Human domination over the earth and non-human species* – Unlike the other two beliefs, this belief is empowering but not necessarily in ways that are consistent with humanity's perpetual obligations. Specifically, this belief is at odds with perpetual obligations to protect earth life from extinction and to protect the essence of nature. In addition to being ethically questionable, this belief is self-defeating as erosion of the health of life on earth, as we are witnessing in real time, not only harms the lives of current generations but also increases the risk of human extinction.
- *Preference for the afterlife* – Of course, for many, life on earth is of lesser concern than life beyond the earth, life in the afterlife, life in heaven, or life in a similar context beyond our reality. There may indeed be life after death but downplaying the importance of life in our reality is quite problematic. First off, this belief seems to imply that the deity also downplays life in our reality, an assumption that is not readily testable and an assumption that begs the question, why create this first step reality in the first place? Second, this belief not only downplays life in our reality but may also preclude humanity from extended explorations of the meaning and purpose of life, which then may have a bearing on how tenable this belief could be. Third, it seems to be a selfish belief because it could reduce the quality of life for others who do not hold such beliefs. Fourth, it seems like a totally unnecessary belief because one can work to ensure the journey in this realm whilst also believing in the afterlife.

- *The Apocalypse* – Lastly, the futures program set out herein is incompatible with beliefs about a supernaturally caused apocalypse. Planning out thousands of years conveys an optimism about our empowerment and ability to survive. Expecting an apocalypse does not. Of course, the nine categories of extinction events laid out in Chapter 3 is a who's who of apocalypses. The differences are that the optimistic perspective holds that we can anticipate and overcome these risks and that there is not a supernatural power waiting in the shadows to unleash the inevitable and inexorable apocalypse.

Exploratory survey research to probe whether religious perspectives impact beliefs about human extinction has been done.[24] As mentioned in Chapter 3, about 45% of the survey respondents believe that humans will become extinct. On an average though, almost two-thirds of respondents and three-fourths of respondents who self-identify as Christians or Jews, respectively, do not believe humans will go extinct, as opposed to 40% who self-identify as secular or non-religious. I hypothesize that the former groups do not believe humans will become extinct because humans will live on in the afterlife.

I understand that both religious tenet and congregation help individuals deal with the trials and tribulations of their lives. I believe that myth and archetype are the core building blocks of religion. They provide comfort. Providing comfort is different, though, from being professing to be the last word on the meaning and purpose of life. The program presented earlier is not necessarily inconsistent with beliefs about the meaning of life or the purpose of life. Moreover, the very long-term perspectives espoused herein might lead some to question whether humanity has existed long enough to be confident that such questions have been well and finally answered.

For the sake of argument, let's say that modern Homo sapiens have possessed the ability to ponder these questions for over 100,000 years. Of course, it has only been relatively recently that we have come to understand the structure of our solar system and universe, the natural causes of apocalyptic events, and characteristics of deep futures. We can appreciate a multiplicity of religious thought. The additional knowledge humanity has gained about reality ought to have some influence on our thinking about the meaning and purpose of life. This additional knowledge should also be humbling, in that we can only be in awe of what more we will learn over the next several thousands of years, that is, if we survive.

During this upcoming journey, our understanding of the meaning and purpose of life may change several times, hopefully in ways that benefit from additional years of existence. For example, some believe that advances in science in general and cosmology in particular have diminished or even demeaned the meaning of human life. How can our lives be important if we live on an unspectacular planet that revolves around a non-descript sun on an outer arm of a typical galaxy, one of hundreds of billions of galaxies in the universe?[25] Well, our lives could be important if the only life in the universe exists on our planet. Our lives could be extraordinarily important if we do learn, at some later date, that we have a role in propagating and protecting life in our universe and ensuring that life survives the death not only of our sun but our universe as well. Of course, these are total speculations but do suggest that human confidence in the meaning and purpose of life, as expressed by religion, is so much hubris.

In summary, aspects of accepted religious tenets are inconsistent with anticipation and sustainability efforts to ensure the journey and meet our perpetual obligations. At the extreme, they are dangerous to our survival in this realm. On the other hand, much of religious dogma is not inconsistent. It is not inconsistent with a belief in a god or gods, or reincarnation, or reverence for all life on earth. It is not inconsistent with imagining better futures. It is not inconsistent with sacrifice for the benefit of others, including future generations. Unlike the preceding two sections, I do not have any recommendations or insights into religion through the Singularity. I only hope that religious tenet can provide succor to adherents while also finding importance in the duty to protect earth life's journey and meeting perpetual obligations.

Cultural diversity

Cultural identity is complex. It is multifactorial. Nationality, wealth, and religion are three components of identity that are typically quite important to people. So are gender, race, and ethnicity. Let's also include political affiliation, caste and class, and profession in the mix. How about taste in music, food, and dress? One's physical attributes beyond gender? Strong affiliations with sports teams or one's city or neighborhood? Or place of worship or place of employment or school or gang? Or perceived intelligence and educational attainment?

Humans are social animals and are hardwired to find psychological satisfaction from membership in groups. Unfortunately, this characteristic of humanity is often counterproductive when groups fear the 'other' and/or when some groups declare their groups superior to others. It is counterproductive when traditionally enforced identities are exclusionary and discriminatory and do not offer individuals sources of identity that they feel pertain to their essence. The best-case scenarios in these situations are manageable tensions. The worst include civil wars, ethnic cleansing, and even world wars. Violence on a wide enough scale could lead to human extinction events, such as nuclear war. The inability of living beings to negotiate these risks when technologies of mass destruction proliferate is assumed to be a main explanation for Fermi's paradox.

The program of ensuring the journey and meeting our perpetual obligations adds a new, major component to cultural identity and an intriguing perspective. The new component to identity is this: every one of every generation belongs to the group called *humanity*. This is not a new idea by any means. Many have argued that discrimination by race is nonsensical and that we can all be traced back to a single ancestor, referred to as Eve.[26] Given continued tensions and violence around the world attributable to racial intolerance, one can conclude that these arguments have not been completely successful.

In this case, though, identifying as human comes with explanations and obligations beyond genetics. The explanations set out why it is important for everyone to identify as human (see Chapter 5) and what this commitment then entails in the form of perpetual obligations (see Chapter 6).

I believe that a very long-term perspective will also be useful to establish this new sense of identify also to eventually reduce tensions and violence attributable to cultural identity. The perspective is simply stated: over the course of the next 1000 years, certainly over the course of the next tens of thousands of years, most every descriptor of identity will change. Virtually no current descriptors of identity will survive into the distant future. Given this perspective, it is virtually impossible to argue that any aspect of cultural identity is best and that intolerance can be logically supported in any fashion.

Most certainly food, dress, and music will change. In fact, the fusion of cultures across the globe is driving cultural creativity and diversity in these areas. Though regrettably many traditional languages are dying out, language per se continues to evolve. The first section of this chapter argues that nation states will necessarily come and go, along with those identities. We are already seeing a proliferation of new nation states across the earth, now numbering about 200.[27] Could this number exceed 1000 as we pass through the cultural Singularity? Culture could change along with the adoption of sophisticated self-sufficiency practices (see the scenario at the end of this chapter). Changes in religious tenets toward futures in this reality could change religious identities. Though genetics has been used as an argument against intolerance based on race, genetic engineering could be used to create a very broad range of new identities. Some have speculated that nanotech could even be used to alter skin color.[28] In Chapter 2, it is anticipated that over time differences between race and traditional ethnicities may diminish as people migrate and reproduce and as cultural themes migrate, fuse, evolve, and diverge *ad infinitum*.

To foster human identity and rationally address the futures of cultural identity, the solution recommended here is to nurture a profound increase in cultural and social diversity. As they say, let 1000 flowers bloom, or in this case, tens of thousands and maybe even millions of flowers bloom into the distant future. Let cultural diversity be the engine of creativity and energy in societies across the world, not the source of sclerosis and violence. Let cultural diversity emerge as the driving force for meeting perpetual obligation # 8, maintain options. By fostering cultural change in organic ways, we can grow and learn what works and doesn't work in peaceful settings.

Imagine:

- Cultural diversity co-evolving with exponential technology change;
- Cultural diversity co-evolving with fundamental changes in economies, such as movements toward sustainable self-sufficiency (see the Transformative Scenario at the end of this chapter);
- Evolving cultural diversity as the engine behind changes in governance and the composition of nations and states;
- Cultural diversity co-evolving with religious tenets that value our reality;
- Cultural diversity co-evolving with the journey and obligations to future generations, with cultural identification with all life over all time;
- Cultural diversity and non-violent transitions in culture reducing fossilized views of the 'other';
- Cultural diversity and change to reduce destructive alienation because everyone can find places where they feel comfortable and accepted;
- Cultural diversity resulting in new scripts, roles, and identities for everyday living.[29]

On the other side of the cultural Singularity, there are no 'others' to be afraid of and no cultural identities could ever be possibly argued as being superior to others. Everyone belongs to group humanity and has a unique set of cultural characteristics. Cultural identity is fluid, not rigid. People will not have to fight, literally, for the right to their identities. Anticipation will be a social norm.[30]

Human psychology

A major theme of this book is that to truly become more futures-oriented and truly care about future generations, the journey requires passing through the socio-cultural Singularity. Earlier, the Singularity is addressed from the perspectives of governance, economics, religion, and culture. In this section, the Singularity is tackled from the perspective of human psychology. This final section addresses the challenges this program presents to the mental well-being of individuals and an approach to help individuals meet the challenges.

Challenges to mental well-being

The program detailed throughout this book and especially in this chapter presents several challenges to individual mental well-being. One challenge is to the core and stability of one's self-identity. A second challenge is to manage the cognitive dissonance that will arise when one's values and self-interest may seem to conflict with humanity's needs to meet perpetual obligations.

Challenge to self-identity – Positive self-identity is a key to good psychological health. As noted earlier, culture provides ample sources of self-identification such as with respect to gender, race, and ethnicity. Self-identity is often conceptualized as what one 'is' versus what one 'is not'. I am an adherent of this religion but not that 'other' religion. I am an adherent of this religion but that 'other person' is an adherent of another religion. Thus, self-identity can be generated through the definitions of differences between individuals and through the identification of the 'other'. In many ways, it does not seem to matter how important or superficial the differences may be.

One could argue that it is hard for people to identify as *group human* because there is no 'other' in this equation. Without an 'other', it can be argued, it is impossible to feel different and special because we are all members of the same group human. If this is the case, then individuals would have little recourse but to rely on old notions of self-identity.

However, given the context of this book, one could present a counterargument that the 'other' from the perspective of group human is *nothingness.* I am human or I am nothing. Without other humans, I do not exist. I am part of group humanity, which is a perpetual group. The 'other' can be understood and felt to be *the void.*

I admit that this is written over dramatically, but the point is that if individuals self-identify as humans then they will advocate for group humanity. It is in their self-interest as they are human and the alternative is nothingness, which offers no possibility for succor and feelings of belongingness. I believe that if group humanity were an intrinsic part of self-identity, then other aspects of self-identity could be more fluid in a positive way. Cultural diversity could more easily flourish if individuals were able to healthily transition self-identities. Fluidity of identities and an explosion of cultural diversity could be expected to lead to increased tolerance for those with different identities, because we could transition to those identities in the future, too.

If individuals can come to value being a part of group humanity, then the achievement of all of the goals presented to this point in the book becomes considerably more probable.

Managing cognitive dissonance – It was argued in Chapter 5 that caring for future generations could, if truth be told, provide more psychological benefits than achievable by not caring or even thinking about the needs of future generations. Caring for future generations and being a positive contributing member of group humanity suggests that individuals ought not to do too many things that can be anticipated to harm future generations. Maybe individuals will come to feel anxiety and guilt about behaving in ways that could harm future generations or not actively and positively contributing to meeting our perpetual obligations.

It is straightforward to see how individuals could constantly experience a sense of cognitive dissonance[31] when confronted with the futures program set out earlier. It will be difficult for individuals to balance individual needs with our personal obligations to group humanity. On the other side of Singularity, our conscience will not allow us to be deceived that a greater good will be achieved if everyone simply maximizes their economic utility and votes only in their blinded self-interest. The old economic and political systems will be seen as neither sustainable nor fulfilling. Thus, individuals will need to address head-on that their behaviors will serve dual purposes to meet their own needs in their daily lives and to help meet humanity's perpetual obligations. We will need to learn how to simultaneously achieve personal self-actualization and contribute to humanity's journey of self-actualization. We will need to acknowledge the real potential for cognitive dissonance, embrace it, and seek assistance when needed to deal with it.

Adaptive work

It is tempting to conclude that all that is needed to surmount issues related to self-identity and cognitive dissonance is more education. It is even more tempting to assume that the world's activities to meet Perpetual Obligation #12, *Fully inform current generations about futures and obligations*, could meet these challenges. Unfortunately, something more rigorous and intensive than typical educational experiences are needed because serious mental and emotional barriers to cognitive change must be overcome.

Ronald Heifetz introduces the concept of *adaptive work* in his book *Leadership Without Easy Answers.*[32] Adaptive work is meant to capture the cognitive effort needed to change one's mind, or in this case, to become more authentically futures-oriented. It captures the amount of energy needed to overcome the stress and anxiety that people feel when their deeply held beliefs are challenged and when they may be faced with modifying their views and opinions. People may literally feel psychological pain when faced with cognitive change. Adaptive work is needed to cope and work through the pain to arrive at a better place, mentally.

I believe that adaptive work is ultimately needed, along with all of the other ideas proposed earlier, for humanity to become sufficiently futures-oriented to support the journey and vigorously meet obligations to future generations. The question for me is whether this goal is achievable from a cognitive perspective. Is there enough plasticity of mental constructs in enough people's minds to effectuate the amount of change described herein? Can individuals deal with changes to their self-identities of culture, race, ethnicity, nationality, and gender or will they resist change? In some ways, it seems like a hero's journey to overcome psychological pain to effect personal change of such magnitude.

Heifetz believes that effective leaders can facilitate successful adaptive work. I am sure not everyone would agree. For example, we could be facing a situation described by Thomas Kuhn in his book *The Structure of Scientific Revolutions.*[33] Kuhn tracks major revolutions in scientific thought, such as whether the earth revolves around the sun and matter is made up of atoms. Older scientists resisted these revolutions in thought, maybe because they lacked cognitive plasticity, felt threatened professionally, or both. Revolutions in scientific thought only became established after the older generations died out. I choose to be optimistic that adaptive work is possible, but I do not want to underestimate the challenge to those working to effectuate cognitive change to more futures-oriented societies.

Here are approaches to consider guiding adaptive work efforts: facilitated cognitive change; immersion; and slow thinking.

Facilitated cognitive change – Heifetz emphasizes that leaders, be they national leaders or grassroots leaders, and in our case also including national anticipatory institutions, need to feel responsibility for recognizing when adaptive work is needed and leading processes that facilitate adaptive work. Since individuals will naturally be resistant to adaptive work, leaders must encourage and even push individuals to perform adaptive work. Heifetz also emphasizes that doing this can put leaders at risk of assassination, metaphorically or literally. He cites as an example of an adaptive work process Lyndon Johnson's pushing American society toward the Civil Rights Act.

Facilitated change can also be led at the community level. Another example cited by Heifetz is an adaptive work process led by the Northwest regional office of the U.S. Environmental Protection Agency (EPA). EPA took on the task of leading the community of Tacoma, Washington through a discussion about why a local manufacturing plant that was emitting asbestos into the environment had to be shuttered. The locals were not happy about losing jobs and decidedly did not value having the federal government intervene in local business decisions. Through a challenging adaptive work process, the participants came to understand that this decision was the best for the future of their health and that the plant could be repurposed in ways to better ensure the future health of the local economy.

Adaptive work processes around futures issues could be held in many ways at the local level. Facilitated discussions could be held in local libraries, churches, and townhall meetings and all those great good places that represent the heart of a community.[34] I can also imagine analogues to Socrates Cafés that address the questions put forward in Chapter 5.

Individuals could also be guided through adaptive work processes by qualified therapists. Just like therapists help their clients deal with the overwhelming loss of loved ones and other enormous difficulties that life presents us, therapists can help individuals change their minds to ease their anxieties about becoming more futures-oriented. The therapy can be custom-tailored and focused on these types of points: caring for future generations can provide increased feelings of group belongingness, as now each and every individual could feel more of a part of the whole of humanity; such caring could increase feelings of unconditional love; one can imagine feeling unconditional love from future generations in gratitude; this is a hero's journey of sacrifice without the guarantee of direct reward; this is the essence of goodness; helping to achieve perpetual obligations can add to the self-satisfaction of achieving goals for oneself, family, community, and nation; and meeting perpetual obligations should help improve sustainability of production and societies, which should reduce stress over having resources and opportunities for everyday survival.

Additionally, many people become quite anxious when asked to think about the future even within limited time frames. Therapists can help their clients overcome these anxieties. Therapists can also help their clients deal with self-identity issues. They can help clients begin to identify with group human and also to healthily transition identities as cultural diversity flourishes.

Immersion – There is no lack of visions of the future. Utopian and dystopian futures are captured in numerous books, graphic novels, movies, and television shows. I imagine that science fiction literacy that gained through exposure to a curated range of these resources could help prepare individuals to tackle real-life futures issues. This type of immersion could be especially valuable to individuals who report not being able to imagine futures very well. Research suggests that the average individuals' ability to imagine the future typically goes dark around ten years into the future.[35] In other words, no image of the future comes to mind at all past this limited time period. Well-facilitated science fiction literacy programs could help these individuals be more receptive to thinking about much longer time frames.

Approaches to using science fiction for immersion should be complemented by immersing individuals in situations where futures are tied to real-life issues. For example, in his book *Resolving Environmental Conflicts*, Chris Maser describes a process he refers to as transformative facilitation.[36] Applied to conflicts surrounding forest management, the first step of the process is to bring all of the stakeholders into a real forest that is the source of conflict. Once there, the facilitator immerses the stakeholders in all things forest, from examining life in the soils and trees to the succession of the composition of forest ecologies over time. Stakeholders are encouraged to share their own views and values about the forest while in situ. Feeling the veritas of differing views, this exercise invariably leads participants to willingly tackle the challenges of adaptive work.

While it is harder to immerse participants in the future, it is not impossible. Early in this century, researchers from the Hawaii Research Center for Future Studies built physical representations of four different Hawaii futures for the year 2050.[37] They invited policy makers and the general public to walk through the exhibits and experience different potential futures for Hawaii. This experience had a positive experience on efforts to plan for the future of Hawaii.

A third approach is to actively involve individuals in futuring. I am optimistic that most everyone can be taught to be more effective and less anxious futures thinkers, in the same way that Tom and David Kelly believe that everyone has the capacity to be creative thinkers.[38] The key to their thinking is simple: run individuals through a series of engaging and effective training exercises. One exercise entails simply giving participants a pen and a paper which has 30 circles and asking them to turn all of the circles into recognizable objects in three minutes. Katie King and Julia West have assembled a similar set of exercises to help individuals think more effectively about the future, such as the exercise presented in Chapter 2.[39]

Beyond participating in short exercises, individuals can be encouraged to become scenario writers themselves. The scenarios could relate to community futures or the future of one's organization. As noted in Chapter 4, it is essential that humanity have a comprehensive and diverse set of SCES available to assess extinction risks. While preparing these scenarios represents a challenge, individuals could join groups of scenario writers to help develop themes and story lines. Lastly, I believe that one part of this solution is to build futures literacy into K-16 education.[40]

Slow thinking – A necessary first step is to become comfortable with thinking about the future. A second step is to improve one's own decision-making with respect to extended future time frames. These decisions can be about one's own self or about humanity. The recommendation here is to train individuals to take their time when making decisions or judgments that involve extended time frames. Using the framework proposed by Daniel Kahneman,[41] the goal is to train people to think *slowly* about these decisions. Over eons, humans have developed decision-making heuristics that can be employed quite quickly to make effective decisions in time-constrained and dangerous situations. Unfortunately, fast thinking is plagued by biases and errors in judgment as seen from the perspective of more considered decision-making. Many of these biases and errors relate to how fast decision-making often does not adequately consider uncertainties. This is an especially problematic situation because almost all decisions involving extended time periods carry with them large amounts of uncertainty. One challenge is how to positively behave in the face of the great uncertainties posed by considering deep futures.

Slow thinking requires focus and effort. It requires cognitive energy. As such, slow thinking is hard and cognitively challenging. Many will not want to engage in slow thinking, which presents another opportunity for adaptive work. This aspect of adaptive work is focused on bringing that extra layer of the journey and perpetual obligations in the consciousness of decision-making. The goal is to help people become less self-absorbed, though it is not the goal to make everyone idealized altruists.

Transformative scenario

It is argued earlier that human society both needs to and the will pass through a Singularity if it is to avoid extinction. I anticipate that the process will take centuries, perhaps by the year 3000 or so. It needs to be noted that the Singularity is not something that will happen to humanity as imposed by some force external to humanity. The Singularity will be created by an accumulation of a gigantic number of decisions made by humanity over time. As such, it is very appropriate to refer to Alan Kay's very famous quote that "the best way to predict the future is to invent it."[42]

Thus, in the spirit of this futures-oriented program, humanity needs to generate a large number of Transformative Scenarios (TSs) for consideration and consultation. Each TS needs to address how the world might look differently with respect to governance, economics, culture, religion, and individual psychology. The TSs will also need to depict how perpetual obligations are being met. It is preferred that the TSs be crafted as stories from the future.

This chapter concludes with an example of a Transformative Scenario. It is called Arachnophobia. It makes progress in reimaging the world in each of the areas just mentioned. However, it still feels like only a partial first step toward really imaging the Singularity.

Arachnophobia[43]

The steaming cup of Appalachian coffee did not disappoint – smooth to the taste yet bioengineered to provide a quick, forceful jolt. Outside, an orchestra of a dozen synthetic species of songbirds was performing an intricate and beautifully coordinated melody. The creatures inhabiting the house were going about their own business, weaving, transporting, digesting, and excreting. A slight vanilla smell permeated the air inside and out. Kate saw her reflection in the window and was again pleased to not look her age. But, with a big sigh, she acknowledged that this delicious, peaceful morning was at its end. Her husband of over a century was noisily muttering as he was making his way to the kitchen.

"Do I have to?" lamented Bob, as his flailing arms almost crushed a beetle crawling up the wall. A smile briefly crossed his face when he saw that a fresh and steaming cup of hot chocolate with community-grown cocoa was waiting for him, along with a plate of newly printed old-fashioned, double-stuffed Oreos.

"Good morning, honey. And, yes, you do," replied Kate, his wife, he hoped, for another century.

"I really have to participate in a Global News Network (GNN) documentary about me? About the space elevator and Probes Project? Our community? Our home? Why? Oh, and I am sorry I didn't say good morning. Good morning!" Kate thought that Bob actually turned a whiter shade of pale as he realized his faux pas. After so many years, why can't he learn to say "good morning" first, they both thought!

Kate's adopted Southern charm kicked in. "Bless your heart, Bob. Thanks to your advocacy for resettlements and now the Probes Project, you have unwittingly become one of the most famous people in the world. And, because you are a dedicated, successful, and, frankly, a paranoid introvert, there is almost nothing in the cloud about you. I know you don't keep track of the collective human consciousness, but I sense the Noosphere daily. Many people simply do not believe you exist. Some believe you are a ghost. Some believe you are a hoax, a fake persona created by the transnational backers of the Probes Project. Your anonymity will be protected, I promise: Your face will never be completely shown and your voice will be disguised through a translation system that will make you sound like a *Muppets* character or a character from that still scandalous animation series *Family Guy*. Sweetie, the world needs to hear your ideas directly and meet you, warts, spiders, and all."

A few hours south of Bob and Kate's home in Southern Appalachia, a documentary team from the GNN had gathered in Atlanta. "Bob really exists!" exclaimed Janet, the documentary team's young and enthusiastic intern. Among her favorite jobs was keeping track of the questions the documentary team planned to ask.

"Yes," replied Melissa, the documentary team's lead. "The GNN upper management was mysteriously contacted and then they somehow arranged to do an exclusive documentary on him. This could be huge for us."

Dan was worried. He was in charge of producing the real-time streaming component of the documentary. He was having a hard time imaging someone he could not picture, much less fathom. "Nobody knows anything about him. He seems to be the only person on the planet with no cloud persona. Where does he live? Where was he born? How old is he? Is he really a he?"

Melissa raised her hands to calm Dan down. "We caught a break. We got a tip from Bob's Decision Day nemesis, the equally mythical Anti-Bob, that he actually lives in an innovative DNA Community located in the midst of several resettlement communities in the Southern Appalachians. He has been in our backyard all this time." Melissa was clearly caught up in the intrigue of the situation.

Decision Day. Bob and Anti-Bob. The team was awestruck. Decision Day was the day the entire world voted as one to decide on whether to cooperate on a global resettlement plan to move tens of millions of people from the areas hardest hit by catastrophic climate change. The vote was extraordinarily contentious. Regions to receive immigrants from around the world were particularly worried. Somehow, Bob's and Anti-Bob's voices rose above the cacophony. Their back-and-forth discussions – which were simultaneously bitter, well thought-out, and considered – clarified the issues. Bob's passion, vision, and can-do attitude won the day in this worldwide vote. The decision directly led to numerous resettlements in places less impacted by catastrophic climate change like Southern Appalachia. Bob and Anti-Bob were also key in Decision Day 2, when the world voted a second time, this time to approve a global effort known as the Probes Project, to find earth life a suitable home beyond the solar system.

The resettlement process in the Appalachians was particularly tense. Those whose ancestors came to the region centuries ago were still fiercely independent and even after a century had not forgotten the government's taking of their land for dams for the Tennessee Valley Authority, the Great Smokey Mountains National Park, nor Oak Ridge reservation's role in the Manhattan Project. Even before the resettlement process began, there were threats of violence. The bombing of the Bangladeshi resettlement, those displaced from their homes by sea-level rise, was tragic but not a particular surprise. Melissa and her crew had covered that story nine years ago.

"Remember the resettlement specialist, Kate, who was on the scene, in charge, and successfully calmed everyone down? Her work paved the way for the peaceful resettlements throughout the United States," Melissa continued. "Well, I contacted her to see if she knows anything about Bob and she does. She's his wife!" Melissa was clearly enjoying telling this story.

The fourth person on the team, Alice, was contemplating Melissa's story in the context of the documentary. Who are the subjects of the documentary going to be? What will the viewers want to see with respect to the daily lives of the documentary's subjects? Was this going to be a drama or a light-hearted farce or a trip down the proverbial rabbit hole? She was leaning toward a drama documentary, completely unaware that she was about to produce a modern version of *Alice in Wonderland.*

John, as usual, was represented by his holographic image. Melissa hated not knowing exactly where John was! He would be responsible for operating the fixed cameras and microphones and tasking the autonomous quad-propeller drones and the robotic tracking bees to follow promising story lines in real time. The holo John was rapidly churning through seemingly every color combination known to humankind. "She is married to Bob? She never said a word! Of course, who knew to ask her about him back then? And, Melissa, my little holographic image of you, floating I tell you not where, suggests there is one more piece to the puzzle. Let me guess – Kate's assistant, the one who was helping her that evening, the one who was pregnant, the one who gave birth right there in the Bangladeshi community center that night and, really, helped solidify the peace, she was her daughter, right?"

Melissa smiled. "You're almost right, John. Kate's assistant was her granddaughter. The baby is Kate and Bob's first great-granddaughter, Roberta." Melissa briefly wondered if Kate had somehow orchestrated the birth, which totally refocused discussions from intolerance to the celebration of life, in anticipation of some violent act. No, that seems quite implausible, thought Melissa. So she continued her report. "According to this note from Kate, they have taken to calling her Bob2, although it doesn't say why. Janet, please note that we need to ask this question. Anyway, Kate goes on to say that her husband was born in St. Louis and had a conventional suburban upbringing."

"That is so cool. We learned about the suburbs in our history class," Janet chimed in.

"Wait, how old is Bob?" Dan asked.

"We do not know for sure," Melissa answered. "He was one of the first to receive the diagnosis of futures vertigo after a major FV seizure while in college. As you know, FV was a mid-twenty-first-century affliction, brought on by the combination of catastrophic climate change and the exponential change and convergence of advanced technologies. Our generation has no problems thinking long term, but back then, many people suffered tremendous mental distress when they tried. Anyway, we can assume he was probably around 20 years old in 2050, so he's approximately 130–150 years old now. His FV seizure turned his life around. He got a degree in sustainable development and then advanced degrees in synthetic biology, engineering ecology, and human-environment systems design. He was one of the first re-environmentalization specialists to successfully integrate wildlife into human settlement in order to re-expand ecosystem ranges and to help save endangered species. Kate says he is known amongst his peers in the Appalachians as Spiderman for his success in saving many species of spiders indigenous to the region." John thought he saw Melissa's holographic image flinch as she read that last bit.

"Bob is head of the life discovery team of the Probes Project. His position is an acknowledgment of his extensive knowledge of life gained from his innovative re-environmentalization efforts, a reward for his role in tipping Decision Day 2 vote in favor of the Probes Project and a result of the fact that the Probes Project is based at the old United States Department of Energy site in Oak Ridge, Tennessee. The first U.S. space elevator will be located there, as will the United States' manufacturing facility for the 10,000 space probes that will be launched next year to explore possible new homes from humanity. We will tour this site as part of the documentary," said Melissa.

Kate continued prepping Bob for the day's upcoming visit. From many years of experience, telling Bob exactly what he needed to do was the only way to relieve some of his anxiety and reduce the probability that anything would go seriously wrong. The pained expression on Bob's face privately caused Kate no end of guilt. She hated her secretive life and manipulations, despite their obvious successes and benefits to Bob and humanity. It was unfair to focus attention on Bob and away from herself, but it had to be done. Maybe one day she would confess everything to Bob, to being Anti-Bob and more, but not today.

"Three things, dear. First, take the documentary crew on a tour of the space elevator site and the Probes Project manufacturing facility. Second, take the crew on a tour of our community and then our home. Lastly, please try to be sensitive to the fact that none of them has been to a community like ours nor experienced a home like ours. As you know firsthand, people's reactions to our community and house can be unpredictable, intense, and maybe even painful. Remember when your brother, George, visited for the first time? He described the experience as falling through the proverbial Kurzweilian singularity. So you must keep an eye on them, understood?" Bob nodded in agreement.

"The lead reporter is Melissa. She was also the lead GNN reporter for the crew that covered the African resettlement bombing. She was there and reporting live when Bob2 was born. She is always well prepared with questions but is high-strung. So, please just try to stay calm. Focus your legendary mind on answering her questions as best as possible. She will probably have four other crew members with her. Again, let me emphasize, try not to let them out of your sight. And, Bob, please have our beloved family AI, TES, inform me when ya'll get to our house. I would like to greet the crew at the front door." Bob and Kate assumed, rightly, that TES was listening to the conversation. Bob assumed, wrongly, that TES would inform Kate when the guests arrived, even though he had not provided explicit instructions.

"OK, Kate, sounds straightforward enough, though I feel quite nervous and uncomfortable about this. You know I am not good around strangers. And there is so much I need to do to equip the probes with the ability to discover life on other planets. I have some more testing areas to set up in some caves and abandoned quarries."

"Yes, Bob, I know you are busy, but really, taking out a couple of days of your life for this documentary will not set this multi-millennial project back too much," replied Kate.

———

Melissa was just about finished. "Kate concludes her note with a few additional details. Bob is on the tall side, thin. He is an introvert but can be quite engaging when you get to know him. Despite rumors floating in the Noosphere from the Decision Day episodes, he is not a Peter Wiggins megalomaniac, nor is he an Ender Wiggins hero type. He is just Bob, who, I must warn you, has the distinctive focus of someone who has undergone neurocognitive restructuring."

The team, including John's floating hologram, just stared at Melissa, not only because they all thought that neurocognitive restructuring was a myth and had never met anyone who had had their brain's memory system cleaned up, but also because the procedure was only done on people over

120 years old. Melissa waved her hand to break the spell that had fallen over the team. "We will be fine as long as we stick together."

TES woke Bob up early on the big day. He and his AI had grown quite close during the lead-up to the first Decision Day. Bob thought of The Everything System as a she, and the AI did not mind using a woman's voice with him. She helped him craft his responses to Anti-Bob's challenges to the resettlement plan. Bob still did not understand why he was singled out or how this Anti-Bob person knew enough about him to call him out before everyone in the world. Somehow, though, his essential anonymity was maintained even as he became a global celebrity. Things did work out for the best. At his urging, and from the contributions of millions of others around the world, humanity voted on that day to embark upon the resettlement plan and then a few years later to authorize the Probes Project.

Both Bob and TES knew that this day was going to be difficult. He was being forced to leave his cocoon, to deal with strangers face-to-face. Kate said that if the documentary was well received, this would be the last time he would have to deal with so many strangers at once. This is because Kate and others could do all of the follow-up interviews once the world knew Bob really existed and that he really was quite busy! Kate gave him a kiss and a hug, as did his great-granddaughter, Roberta, who was spending a few days at their house with several cousins while the various sets of parents were visiting other DNA communities around the country.

The tour of the probe manufacturing facility went well. Bob answered the crew's many technical questions. Dan and John were particularly interested in Bob's strategies for outfitting and programming the probes to detect life. For example, Bob was using not only the DNA Community but also his own home as a test bed for the probes' life-detecting systems. Scattered throughout the surface waters and streams were nonliving chemical reactions and very rare biological organisms, like the ancient bacterial mats. Many things in the community moved, from inert wind chimes and propellers to robotic bees. Even caves were outfitted with a mixture of organic and nonorganic systems. Soon a probe would be launched in a geostationary orbit above the community and tasked with determining what was alive and what was not. Everything was designed to train and test the probe's life-detecting algorithms. Bob was even happy to share his thoughts about the newly evolving field of terraforming, even though that knowledge would only be applied outside of the solar system tens of thousands of years in the future.

As he saw how impressed the crew was with the space elevator site and the probe manufacturing facility, Bob got over his nervousness. They soaked up both the technical information and the history of the place. Only Janet, who had a knack for history, remembered that the site was home to the first operational nuclear reactor and the facility that produced the radioactive materials for the atomic bombs dropped on Japan during World War II. However, the crew was acutely aware of the global resettlement program. Hundreds of millions of people relocated to more sustainable homes, including millions from around the world to the relatively sustainable southeastern United States. Bob was patient and persuasive as he explained that the resettlement program led to concerns that life on earth could not be sustained indefinitely, that new homes needed to be found elsewhere in the universe, and that the Probes Project was an essential step toward realizing this goal.

The crew was surprised to learn that Bob's community, the DNA Community, coevolved with the numerous resettlements in the region. The DNA Community contained like-minded people, in this case with a shared fervor for all things biological rather than a shared ethnic or national heritage. Like resettlement communities, the population of his community was about 200, following Dunbar's tenets about the optimal group size. Still, everyone was a bit apprehensive as they hopped in the autonomous GNN van for their tour of the community.

The tour of the community went well, too. The DNA Community was certainly quite different: everything seemed to be alive. The vanilla scent was quite pleasant. There was no concrete or asphalt. Bob showed them the central agricultural facility, which they affectionately called the Sprawl Farm. A complex system of fuel cells, photovoltaics, wind, and microhydro generated electricity for the community and also provided food. An equally complex 3D printing facility was busy printing gastronomic delights at the molecular level at one end, medicines at the other, and household objects in the middle, with all of the inputs grown right there or recycled from old products. An energetic carbon-lego assembler was constructing a table. The team began to relax, which infused Melissa with a sense of foreboding.

———

Kate removed her earbuds in time to hear a scream and a thud from the playroom and the sound of the front door closing suddenly. "Oh, Bob! You promised to let me know when you arrived." After taking a few minutes to quiet down from the eventful tour of their home, Bob and Kate sat in the jungle room of their home. "Bob, tell me what happened."

Bob ran his hand through his thinning hair. "OK. Really, I don't know what happened exactly, but I'll tell you what I know. The tour of the space elevator site and probe manufacturing facility went well. They loved our community and the Sprawl Farm. I gave them a taste of that new chocolate pastry that was just printed – you know, the one with the mood-calming additive. Then we headed over here. They came in a big van but left their equipment inside because this was just an exploratory visit. They asked a lot of questions as we walked to our house. Why was this called the DNA Community? Was the community truly self-sufficient? Did we live with wild animals? Was our house made of mud? Why did 200 people live in the enclave, and why was 200 a magic number? Did our children run around naked? Really, they had a lot of misconceptions about our community and house, so I was glad that this first visit was just to scope things out."

Kate interrupted. "Did they seem apprehensive? Had they had any desensitivity training? You know that even the most educated people have trouble dealing with their fears of animals and insects and spiders? Did their questions raise any red flags with you that this visit might not go very well?" Bob squirmed a bit and worked on relieving his head of some more hair. "Maybe I should not have fed them the pastry. Seemed to really calm them down. So, they seemed fine but in, ah, maybe in retrospect."

Kate could not contain her exasperation, "Oh, Bob, how many times have we been over this?"

About the same time, Melissa had finally woken up in the local clinic. Her team was around her hospital bed. She went from groggy to focused fury in a matter of seconds. "Where were you all? I told you to stay by me. I told you that we needed to stick together to get through this first trip. What happened to each of you? Did you even make it into the house?"

Janet was dirty and frazzled. "During the walk to the house, Bob was explaining the enclave's composting operations, specially designed creek beds for endangered mussels, fish-farming tanks, the revival of the American chestnut through gene drive technology, and biomass crops. I really wanted to see all this, so I lagged behind the others to walk around the enclave a bit more."

"I wandered around the back of the house and then out behind one of the operations buildings. I could see a huge pile of smoldering dirt. The smells began to overwhelm me. The rotting of the compost pile was apparent. Then, came a very strong fish smell. As I turned the corner, a big front-end loader dumped thousands of fish carcasses on top of another compost pile. I walked up to take a closer look. The woman on the tractor started waving her hands, getting more and more agitated. A moment later, hundreds of weird looking sea gulls swooped down on the pile. They landed on me, knocked me down, and pooped on my hair. I hate birds and they hate me. I totally freaked out. Then a bear came out of the woods, followed by a bobcat and its two kittens. I thought they were going to eat me. Then something that looked like a dinosaur came out of the woods, too, like it ran right out of Jurassic Park. I ran into the woods, through several large spider webs. I can still picture one of the spider's eyes, staring at me in shock! I also managed to roust a nest of very strange-looking bees, which scared the honey out of them and me. I still don't know how I made it back to the van."

"When we got to the house, there were only four of them. I really don't know what happened to the fifth person," Bob reported to Kate. "She happened upon the fish dump at exactly the wrong time," surmised Kate. "Oh," responded Bob is a very weak voice. "Continue, please," directed Kate.

"OK, well, we got to the house. I asked everyone to leave their shoes outside, like always. No one was wearing socks."

"I am glad you noticed that," said Kate, very sarcastically.

"Hey, no need for that! I had everything under control. The Jungle Room was neat, and everything was picked up. The music was already playing. They seemed to really enjoy the music. We had a few minutes to tour and then get out of the room. Everything seemed fine."

———

"OK, Janet. Sounds like you had a harrowing experience. But that is no excuse for wandering off. Anyway, if you had been listening more closely to Bob," Melissa lectured, "you would have known that those weren't seagulls; they were passenger pigeons brought back from extinction and given sea gull genes to crave fish. And yes, that was a dinosaur, a miniature velociraptor, which was also brought back from extinction. And, I do believe that was a nest of robotic bees programmed to pollinate the DNA Community's fields. OK, OK. Back on point. So, what happened next? We were in the front room, the so-called Jungle Room. Bob seemed intent on getting us through the room quickly. Then we headed toward the back of the house." Alice looked the most sheepish at this point, so Melissa targeted her for the next interrogation.

"The room was so interesting! Grass was growing on the floor. It felt so soft. The music was hypnotic. What was it again?" Dan interjected that Bob said it was his 12th chromosome. "Wow, I wonder what my DNA would sound like. Anyway, Bob seemed in a hurry to move us along but I wanted to stay just a second longer. The far wall appeared to hold many interesting artifacts. I wandered over to have a better look."

———

"I had plenty of time to get the visitors through the Jungle Room before the mist started. How was I to know that one of them would linger?"

"Bob, you just need to be more aware of what is going on? Did you ever look back or count heads?"

"No, I guess I was too involved in explaining our wonderful house to Melissa."

"Oh Bob," muttered a clearly exasperated Kate.

———

"What I thought were artifacts were dens and homes for creatures and the living creatures themselves. I reached out to touch one and it moved and then disappeared into the wall. Several others also disappeared. One ran down the wall and bored into the floor. Then it began to rain! Flying insects emerged from the walls. They started to get into my hair. I fell back against the wall. My hand got stuck on some sap. It was dirty and gross. I was struggling to pull my hand out, when a snake crawled across my bare feet. I began screaming hysterically, but because of the noise of the misting and because the doorways to the rest of the house were now closed, no one heard me. The adrenaline rush gave me the strength to pull my hand out of the wall but not the brains to tell me not to put my hand in my hair. For some reason, the front door just opened on its own, and I ran out of the house wet, dirty, and with my hand stuck to my head."

"So, Alice, that explains your new, ugly buzz cut and your dirty, hairy hand," remarked Melissa. "That also explains why Bob rushed most of us through that room while explaining the house's compartments for living creatures to me in excruciating detail."

———

"I was telling Melissa about the house. I didn't want them to be afraid of anything coming out of the Jungle Room during the mist. I think Dan understood; at least I thought he did. He seemed to understand that although the house was constructed of carbon nanomaterials, all the walls were custom-designed to either accept and house or repulse various species. He seemed to understand that this was accomplished by the use of pheromones in the soils, as well as colors, other smells, textures, humidity, temperatures, water and food availability, the sizes of passages, sounds, and even electrical currents. In fact, each room had its own special ecosystem. Then he asked where the bathroom was."

"What, you let him go to the bathroom without any special instructions or warning," exclaimed Kate.

"Like I said, he *seemed* to understand what the house was all about."

———

"Janet and Alice screamed and I fainted. Great! OK, men. What are your stories? Why did you abandon damsels in distress?" Melissa goaded.

Dan was next. "Melissa, I didn't intentionally abandon you. I just went to the bathroom. I really had to go!"

"Yeah, yeah, we all know about you and your bathroom habits. So, what could possibly go wrong in a bathroom?"

———

"I pointed him to the bathroom. I told him we had a composting toilet with a very new design. That he would enjoy the experience. That he should take his time. That after walking through the kitchen, we would head on to the back of the house to visit the playroom."

"In other words, you didn't give him a clue about what the bathroom is like?" questioned Kate. Bob could only shake his head.

"So, anyway, I really had to go. As I was running off, I heard Bob instruct me to try to tell the difference between the beetles and something else. Honestly, I was not paying too much attention at this point. Not to go into too much detail, but I hurried in, closed the door, and settled in. It was crazy but I swear it smelled of vanilla and wintergreen in there," mused Dan. John interjected that the pervasive vanilla smell was probably due to some bioengineering of digestive track microbes to produce sweet-smelling poo. The others mulled this over for a bit before agreeing.

"There were no windows or electric lights. Bioluminescent plants provided the light. Like the Jungle Room, this room seemed to have stuff all over the walls and ceiling. In fact, there were bugs and things everywhere. I am not usually afraid of insects. But, in a small plant along the far wall, there was some interesting activity. Good thing I had on my binocular glasses. I focused them on the scene across from me and began to watch. That was a bad idea. My worst nightmare, being eaten by a female, was playing out in front of my eyes. A big female praying mantis was eating her mate. Chomp, chomp, chomp. I kinda lost track of things at this point." It looked like Melissa was about to eat Dan!

Dan continued, "My skin began to get prickly and I began to feel nauseous. Then I felt something crawl across my foot. Then something else. I let my gaze drop without readjusting the glasses and looked down. Gigantic dung beetles and ants were streaming out of the bottom of the toilet, carrying my, well, you know what up the walls! Some of them even looked robotic! I freaked out but didn't scream, I want you to know! I jumped up, but forgot that my pants were around my ankles. I tripped and fell into the shower area. The shower immediately turned on but the first mist was not water but mites. I know this because my face was smashed against the floor and even though they are small, up-close I knew! They were there to eat my dead skin. Then a heavy shower soaked me and then, well, I passed out. The next thing I knew I was out of the house, with a sensation of being carried out of the house by many very little things, like Gulliver. Oh, my God, I probably was," Dan cringed as he realized what had happened.

"So, you never saw Dan again?"

"That's right. I went back to find him but he wasn't there. But he took a shower! Why would he do that?"

"Oh Bob, I can assure you that he didn't voluntarily take a shower. Anyway, you said you took them through the kitchen? I was in the kitchen and didn't see you."

"We saw you. You had your ear buds on. You don't like my 12th chromosome opus, remember? You were frying beetles and ants for our pet aardvark. There was a plate of goodies on the table. I didn't want to bother you, and yes, I forgot that you wanted TES to let you know we were in the house."

"Oh, my, Bob, I think I know what happened to John."

John couldn't stop laughing. The image of Dan being carried out of the house by a stream of insects with his pants around his ankles was just too funny. Melissa rudely interrupted. "John, what is so funny? Why did you run out of the house? Why do you look green around the gills?"

"OK, OK. I admit it. I am no better than anyone else. But I could have died! After Dan scurried off to the bathroom of horrors, we continued on to the kitchen. Bob's wife, I take it, was in there, at the stove. She had some headphones on and didn't hear us come in. Who would in any case, with floors made of grass? Anyway, Bob didn't want to bother her so he just kept going. Like Alice, I lingered a bit too long in this room. It was fascinating. There were herbs growing in niches in the walls. There were grapevines growing up the walls. There seemed to be tree branches coming through the walls that held apples and oranges and lemons. How did they engineer that? Anyway, it was interesting and I was finally beginning to feel at home in this crazy place. On the table was a plate of fried chicken nuggets that smelled delicious. I picked one up . . . gave it a good look. I have always thought I am not as stupid as I look but, well . . . Anyway, I ate it. Crunchy goodness. Another intoxicating taste, like the pastry. Then I ate another. The plate was stacked, so I had a third and a fourth. I was eying a fifth tasty treat when I noticed some beetles crawling up the wall of the kitchen in front of the stove. The wife reached out, picked one right off the wall, and threw it into the frying pan. A side door miraculously opened. I managed to make it to the edge of the woods before heaving."

"So, now it is just you and Melissa and you still didn't think anything was wrong?" asked Kate.

"No, like I said, I was deeply engaged in conversation with Melissa. She seemed really interested in everything. She was itching a mosquito bite and

I told her that was good because she was probably just injected with a vaccine for the West Nile virus. She had a bit of a headache coming on, so I gave her a medicinal cookie to eat. She said she really wanted to see the playroom because she has two young kids whom she really adores. She is always interested in new play ideas for her kids," said Bob.

"Bob, you of all people, you who has spent a career in re-environmentalization, who is unwittingly internationally known for designing human – nonhuman interfaces, should have remembered that a large fraction of the population has serious phobias. You missed them all this time. I bet that first poor woman suffered from melanophobia, a fear of birds. And the second one, who most certainly ran out of our house screaming, probably suffers from ophidiophobia, a fear of snakes, and probably from spernatophobia, a fear of germs. The guy in the bathroom probably suffers from pmyremecophobia, a fear of ants, which was triggered by the beetles. Didn't it occur to you that Melissa might suffer from arachnophobia?"

"She never mentioned spiders. Never asked why they call me Spiderman. I purposely did not meet them with any spiders on my shoulders. I didn't show them the spiders under the awnings at the side of the house. So, please understand that I really had no idea," protested Bob.

"OK, Boss," said Dan. "What happened to you?"

"I couldn't divert Bob's attention from his monologue. He is clearly a genius. His focus is astounding, as expected. And what he was talking about is genuinely worthy of a documentary. The house and the community are, as you all have said, fascinating. After really preparing ourselves by taking as many weeks of de-sensitivity training, as needed, we are going back, no protests!

"Anyway, by this time, I was apprehensive. Given Bob's nickname, I had been dreaming of spiders every night for the past week and waking up in a sweat each time. Then, the day before yesterday, I saw a small spider on my daughter's shoe. I freaked out and she screamed. I beat it to a pulp with my portable umbrella. It didn't seem like a good sign, given our pending visit to the DNA Community. I really didn't want to see any spiders during our visit, and I really wanted at least one of you to be with me at all times! I had not seen any signs of any spiders, but then again, you all had disappeared."

Melissa continued. "After the kitchen, we headed toward the playroom in back of the house. Bob was talking about how it was hard to live with the black bears. They need a large range and substantial food supplies. If they cannot find their natural foods, they will scavenge for food in dumpsters and garbage cans. This causes lots of problems, for the bears and humans alike. His community has devised paths, which travel over roads, under roads, over rivers, and over railroad tracks, for the bears to follow when they move from one protected area to another. The paths have invisible electric fields to help keep the bears on the paths and moving along. In addition, people are discouraged from hiking the bear paths. They have planted more acorn-bearing oak trees as bear food sources, both in the mountains and in other protected areas. We have personal experience with the fish carcass dump, too, right, Janet? They have developed a black bear educational website, too."

"Out with it, you are stalling," Dan interjected.

"I'm getting there. Bob also talked about what they are doing to deal with coyotes and how they are getting people to actually let bats live in their belfries. The big Appalachian resettlement, which has been managed by his wife, if you didn't know, provided huge opportunities to design new enclaves from the ground up to support re-environmentalization efforts. The newcomers have been receptive to green roofs and homes on stilts with fungus farms underneath. They understood the need to quarantine potentially invasive species. They came here with many fewer phobias than the current residents have," said Melissa.

"Then Bob mentioned his great-granddaughter, Roberta. I remembered the story of her birth, in the Bangladeshi enclave the night of the firebombing. There was a picture of her on the wall, heading to the playroom. She looked so sweet! Bob said that Roberta took after him more than Kate, his wife. Bob hoped that Roberta would eventually head up the Probes Project after his time on this earth, which is why some were calling her Bob2. Just as I asked him why he thought this was a good idea, he opened the door to the playroom. Roberta came running up to me. Covered, covered." Melissa again looked faint.

———

"What, Roberta was covered in spiders? And her cousins were, too? Oh, Bob," Kate cried. TES, who again had been uncharacteristically quiet during this whole discussion, finally blurted out through an image in a painting in the wall, "And I recorded it all! What fun! I hope this material makes it into the documentary."

———

Once Alice saw the footage of the team's initial visit, and similar mishaps from subsequent visits, she scripted the documentary to include TES's material for a humorous rather than dramatic result. Although it took every ounce of persuasive talent, she had to convince Melissa to approve this tack, it turned out to be the right approach: the documentary went viral. Interest in biologically centric human settlements increased, along with the numerous areas of science underlying these communities' designs. More people began incorporating the various ideas from the DNA Community and Bob's home into their lives. Interest and support for the Probes

Project also rebounded nicely. Lastly, and perhaps most importantly, more communities and homes smelled of vanilla!

Notes

1 R. Kurzweil. 2005. *The Singularity Is Near: When Humans Transcend Biology*. Penguin Books, New York.
2 D. Elgin. 1993. *Awakening Earth: Exploring the Evolution of Human Culture and Consciousness*. William Morrow, New York.
3 www.simplypsychology.org/milgram.html
4 www.prisonexp.org
5 https://en.wikipedia.org/wiki/The_End_of_History_and_the_Last_Man book by Francis Fukuyma.
6 The act of waging war to demonstrate which system is best seems to suggest that the systems themselves are fatally flawed.
7 B. E. Tonn and D. Feldman. 1995. Non-Spatial Government. *Futures*, 27, 1, 11–36.
8 B. E. Tonn, D. Stiefel, J. Scheb, C. Glennon, and H. Sharma. 2012. Future of the Courts: Fixed, Flexible, and Improvisational Frameworks. *Futures*, 44, 802–811.
9 B. E. Tonn and D. Stiefel. 2012. The Future of Governance and the Use of Advanced Information Technologies. *Futures*, 44, 812–822.
10 B. E. Tonn. 1996. A Design for Future-Oriented Government. *Futures*, 28, 5, 413–431.
11 Designed for a country the size of the United States.
12 https://en.wikipedia.org/wiki/Dunbar%27s_number
13 That is, if it is a failed state https://fragilestatesindex.org
14 https://en.wikipedia.org/wiki/Social_discount_rate
15 www.fao.org/3/X8955E/x8955e03.htm
16 https://e360.yale.edu/features/ecosystem_services_whats_wrong_with_putting_a_price_on_nature
17 As this is being written, this disconnect is a growing concern in the United States. The downturn in parts of the economy, especially in the areas of travel and entertainment have reduced state and municipal tax revenues. Governments are now faced with making severe cuts to programs even though the human and physical infrastructures are there and the need for government services is more dire.
18 M. Hart. 1999. *Guide to Sustainable Community Indicators*. 2nd ed. Hart Environmental Data, North Andover, MA.
19 Though, sunk costs and opportunity costs could also be perceived as being myopic and short-sighted if the time frames encompassed are short and the costs involved are small picture (e.g., limited to one individual's happiness).
20 https://dschool.stanford.edu/classes/design-for-extreme-affordability
21 https://all3dp.com/2/coolest-3d-printed-cars/
22 I am particularly drawn to the concept of vertical farms. Located in dense urban areas, vertical farms are located in high-rise type buildings where each floor is dedicated to growing crops, animal husbandry, aqua-culture, etc. https://en.wikipedia.org/wiki/Vertical_farming
23 The term deity as used herein is meant to reference a god or gods or other supernatural being or beings that essentially guide the lives of devotees according to the will of the former.
24 B. E. Tonn. 2009. Beliefs about Human Extinction. *Futures*, 41, 10, 766–773.
25 www.space.com/25303-how-many-galaxies-are-in-the-universe.html

26 www.newscientist.com/article/mg22429904-500-found-closest-link-to-eve-our-universal-ancestor/
27 www.worldometers.info/geography/how-many-countries-are-there-in-the-world/
28 www.frontiersin.org/articles/10.3389/fchem.2019.00739/full
29 E. Goffman. 1956. *The Presentation of Self in Everyday Life.* Doubleday, New York.
30 T. Fuller. 2020. Anticipation and the Normative Stance. In R. Poli (Ed.), *Handbook of Anticipation.* Springer International Publishing. Retrieved from https://link.springer.com/referencework/10.1007%2F978-3-319-31737-3
31 "Cognitive dissonance refers to a situation involving conflicting attitudes, beliefs or behaviors. This produces a feeling of mental discomfort leading to an alteration in one of the attitudes, beliefs or behaviors to reduce the discomfort and restore balance." www.simplypsychology.org/cognitive-dissonance.html
32 Ronald Heifetz. 1994. *Leadership without Easy Answers.* Harvard University Press, Cambridge, MA.
33 T. Kuhn. 1962. *The Structure of Scientific Revolutions.* University of Chicago Press, Chicago, IL.
34 R. Oldenburg. 1999. *The Great Good Place: Cafes, Coffee Shops, Bookstores, Bars, Hair Salons, and Other Hangouts at the Heart of a Community.* Marlowe & Company, New York.
35 B. E. Tonn, A. Hemrick, and F. Conrad. 2006. Cognitive Representations of the Future: Survey Results. *Futures,* 38, 810–829.
36 C. Maser. 1996. *Resolving Environmental Conflict, toward a Sustainable Community Development.* St. Lucie Press, Delray Beach, FL.
37 S. Candy, J. Dator, and J. Dunagan. 2006. *Four Futures for Hawaii 2050.* Hawaii Research Center for Future Studies, University of Hawaii at Manoa, Honolulu, Hawaii, August 26. Retrieved from www.researchgate.net/publication/253641086_Four_Futures_for_Hawaii_2050
38 T. Kelly and D. Kelly. 2013. *Creative Confidence: Unleashing the Creative Potential within Us All.* Crown Business, New York.
39 K. King and J. West. 2017. *Futures Thinking Playbook.* Teach the Future, Houston, TX.
40 R. Miller. 2018. *Transforming the Future: Anticipation in the 21st Century.* Routledge, London.
41 D. Kahneman. 2011. *Thinking Fast and Slow.* Farrar, Straus and Giroux, New York.
42 https://en.wikiquote.org/wiki/Alan_Kay
43 This scenario was first published in Flash Forward: A Series of Futuristic, N. Vignettes Savage and A. Street (Eds.). Jenny Stanford Publishing, Singapore, 2016.

10 Futures thoughts

I hope that you have found the journey taken by this book to be worthwhile, maybe even uplifting and exhilarating. By now it should be completely clear that becoming more futures-oriented represents an extraordinary challenge to humanity. I hope that a radically long futures focus on life can bridge political and cultural divides that so much define our myopic battles for power and ideological supremacy. It should also be clear that at least from an intellectual point of view, the challenge can be met, not today or tomorrow but certainly in the not-too-distant future with a lot of hard work.

Recap

Chapter 2 took us through a serendipitous journey through future time. There are many events that we can anticipate with near certainty, such as the death of our sun, the movement of continental plates over geologic time, the eventual decomposition of nuclear wastes into harmless substances, and the exhaustion of fossil fuels in approximately 500 years. There are many events that are anticipatable given current trends and our current knowledge, such as a climate catastrophe and the threats that genetic engineering technologies could pose to the nature of nature and humanity. The exercise highlighted in Exhibit 2.1 was designed to build confidence that thinking about the future is doable, useful, and fun as well.

It cannot be argued that catastrophes of various sorts, including human extinction, are being completely neglected by the current generations. On the other hand, Chapter 3 adds to the discussions by taking a very comprehensive approach to extinction risks. Nine categories of extinction risks are presented. About half are anthropogenic in origin. About half are natural or extraterrestrial, which we probably cannot prevent but can anticipate and plan around. The extinction risk scorecard communicates that humanity is presently not well addressing near-term anthropogenic risks nor any other category of extinction risks very well.

Another contribution of this book is to explore extinction risks that are not simply the result of one extinction-level event. Chapter 4 introduces singular chains of event scenarios. These scenarios are composed of many events, only some of which might be extinction-level events, which at the end humans are unable to successfully navigate to prevent their extinction. A detailed SCES is presented at the end of this chapter. An additional consideration for the development of SCES is the unintended consequences of events, which includes introduction of new technologies and new advancements in science. Yet another consideration is unknown unknowns. It is not just unintended consequences that humans must do a better job of anticipating. We must also be humble in admitting that much (most?) of reality is hidden from us and maybe even new aspects of reality are emerging from the chaos of the world we created. We need to be vigilant. It is recommended that a Whole Earth Monitoring System be developed to capture data that will aid evidential reasoning and decision-making and that might reveal previously unknown unknowns.

Chapter 5 promoted a commitment to the well-being of future generations. A commitment to future generations, the journey, and self-actualization of humanity through time and space should be part and parcel of being human. It is argued that the concept of the journey is a powerful myth (i.e., archetypal story) of humankind. It is argued that caring for future generations can provide an additional path for psychological well-being. It is also argued that rationally if we agree that we should care about our own children and grandchildren, it then follows that we should care about all future generations.

Chapter 6 lays out the framework to move from why we should care about the journey and future generations to how to turn caring into policy. This chapter presents a dozen perpetual obligations that, it is argued, every generation should strive to meet. The twelve can be considered an aggregation and enhancement of previously presented obligations to future generations, such as contained in the UNESCO Declaration of Obligations to Future Generations. Metrics to determine whether each perpetual obligation is being met are proposed. Numerous scenarios, such as SCES, are needed to help estimate the metrics. The five types of proposed themed scenarios are presented in Exhibit 6.2. The scorecard presented in this chapter suggests that humanity is not currently meeting any of the twelve proposed perpetual obligations.

Chapter 7 presents a six-level action framework that suggests levels of effort needed to be expended to meet perpetual obligations depending on the seriousness of the situation. The action levels run from Do Nothing to Manhattan-scale Projects to War Footing. It is suggested that methods that fall under the rubric of imprecise probabilities be used to estimate upper and lower probabilities of human extinction and implement an evidential reasoning scheme for qualitative metrics associated with perpetual obligations. A matrix is presented that links perpetual obligation metrics to each action level. It is suggested that humanity develop 1000-year plans to meet perpetual obligations. Lastly, it is recommended that humanity build the Global Anticipation and Decision Support System to foster international cooperation and to link perpetual obligations to global goals, and national and subnational plans, programs, and policies.

The perpetual obligations and frameworks presented in the previous chapters lay the foundation for futures-oriented policy, but futures-oriented policy institutions are needed to make the actual decisions. Chapter 8 presents recommendations for international and national institutions that are needed to use the frameworks presented in the preceding two chapters to be the conscience of the policy making world. The InterGenerational Panel on Perpetual Obligations will provide humanity with inputs about how well humanity is meeting its perpetual obligations. The World Court of Generations will judge how well humanity is meeting its perpetual obligations and suggest which action levels are needed. National Anticipatory Institutions will make their own judgments. A goal of an international process is to develop the Perpetual Obligations International Protocol. Finally, a small number of perpetual organizations, such as the Stewardship Institution to manage nuclear wastes, are proposed.

In Chapter 9, I argue that five essential aspects of human society need revolutionary changes to help ensure our journey. Imagery of passing through a socio-cultural Singularity is employed to describe the changes required. Discussed in this chapter are changes to approaches to governance and long-term experimentation of political systems. Diversity of culture and socio-cultural change must be tolerated not only to maintain options for future generations but also to allow exploration of cultural change. Such change could manifest itself politically as well as socially. Economic theory and economic realities on the ground must also be open for reassessment and evolution. Religious dogma that presupposes determinism, fatalism, and lack of free will are at odds with this entire program. Cultures should strive to be more diverse, tolerant, and fluid. Our own psychology will need to evolve to find fulfilling identities in a more fluid world and be more futures-thinking competent.

Figure 10.1 presents a final scorecard that combines assessments made throughout the book. Overall, humanity does not score well. There are no passing grades with respect to meeting perpetual obligations, the application of anticipatory methods with respect to extinction-level events or global catastrophic risks. It seems to me that humanity needs to implement Manhattan-scale projects to address threats to human extinction, species extinction, sustainable production, and involuntary environmental risks. Major programs and policy nudges are needed with respect to the balance of the dozen perpetual obligations.

Enormous amounts of work that can be projected to require centuries of effort are needed to create WEMS and GADSS as well as new futures-oriented institutions. Humanity is barely on its way to passing through the socio-cultural Singularity. Progress in all of these areas will work hand in hand with increasing the care we have for future generations. I expect the scorecard will improve over the centuries! I cannot imagine that humanity

Source of Existential Risk/Evaluation Criteria	Societal Status of Anticipation	Societal Status of Planning	Societal Status of Mitigation Efforts	Current Ability to Adapt	Current Estimate of Survivability
I. Anthropogenic	P	C	C	F	C
II. Coupled Human-Env. SYS	C	C	F	F	C
III. Human Reproduction	F	F	F	F	C
IV. Risk to Humanness	F	F	F	F	C
V. Advanced Technology	F	F	F	F	C
VI. Natural Terrestrial	F	F	F	F	F
VII. Solar System	F	F	F	F	F
VIII. Extraterrestrial	F	F	F	F	F
IX. Universe Scale	F	F	F	F	F

Application of Anticipatory Methods	With Respect to Extinction-Level Events	With Respect to Global Catastrophic Risks
Best Practices	F	C
SCES Development	F	F
UUCs Analyses	F	F
UKUKs Analysis	F	F

Reason to Care About Future Generations	Propensity to Support Future Generations
Deeply Held Values	70%
Supporting the Journey	50%
Duty-Bound Contract	10%
Protect Earth Life	10%

Perpetual Obligations	Suggested Action Level
#1. Prevent human extinction	Manhattan Project
#2. Prevent extinction of life on earth	
#4. Bequeath sustainable production	
#5. Reduce involuntary environmental risks	
#10. Support generation of new knowledge	
#11. Finish important business	
#3. Bequeath sustainable societies	Major Program
#6. Preserve essence of nature	Policy Nudges
#7. Preserve essence of humans	
#8. Maintain options	
#9. Preserve human knowledge	
#12. Fully inform current generations	

Recommended Creations	State of Development
WEMS	10%
GADSS	5%
International Institutions	0%
National Anticipatory Institutions	2%
Perpetual Organizations	0%

Aspects of Humanity	Progress Towards Singularity
Political Theory and Organization	1%
Economic Theory and Organization	5%
Religion	20%
Culture	30%
Individual Psychology	5%

Figure 10.1 Final Summary Dashboard

could ever become complacent in dealing with the challenges of meeting its perpetual obligations.

Major research and activity categories

Chapters 2 through 9 set out a blueprint for humanity to follow to become more futures-oriented. Since this is only a blueprint, there are many things to be accomplished. These tasks fall into six major categories:

(1) Ethical Thresholds and Other Metrics

- Establish acceptable risk of human extinction;
- Establish acceptable risk of extinction of all of earth life;
- Establish a threshold of when the essence of nature is lost;
- Establish a threshold of when the essence of human nature is lost;
- Establish rigorous metrics for the other eight perpetual obligations.

(2) Metrics assessment

- Estimate upper and lower probabilities of human and earth life extinction.
- Estimate upper and lower probabilities of death due to involuntary environmental risks.
- Estimate upper and lower probabilities of worldwide production systems being sustainable over a 1000-year time horizon.
- Estimate metrics for the other eight perpetual obligations.

(3) New Programs

- Establish and expand research and development programs to ensure the journey;
- Design, test, and implement approaches to protect human knowledge;
- Establish humanity-wide no-regrets and unfinished business projects;
- Design, test, and implement global futures communications strategies;
- Develop 1000-year plans.

(4) New Institutions

- Establish the InterGenerational Panel on Perpetual Obligations;
- Establish the World Court of Generations;
- Amend constitutions to establish National Anticipatory Institutions;
- Negotiate and ratify the Perpetual Obligations International Protocol;
- Establish Perpetual Organizations.

(5) New Information Systems and Resources

- Design and administer the Whole Earth Monitoring System;
- Design, test, and administer a subcomponent of the WEMS related to the human exposome;
- Design and administer the Global Anticipation and Decision Support System;
- Design and administer systems to support non-spatial governments;
- Implement global process to foster the development and vetting of Singular Chain of Events, Unintended Consequences, Transformative and other themed scenarios.

(6) Change Leading to the Socio-Cultural Singularity

- Design, implement, and evaluate experiments with new systems of governance;
- Foster an environment to promote an Einsteinian revolution in economic theory and self-sufficient production;
- Foster synergies between futures perspectives and religious tenets;
- Foster an environment that encourages cultural diversity;
- Implement empathetic adaptive work programs to ease the transition in the areas of futures literacy, futures decision-making, and self-identity.

Last thoughts

A major theme of this book is that to truly become more futures-oriented, to truly care about future generations and the journey, requires a brand-new layer of consciousness for humanity. The perpetual obligations represent a new layer of what it means to live a purpose-driven life. There will be a new layer of individual identity. There will be new layers upon our understanding of economics and organizations. There will be new layers in our perspectives of governance and political organization. There will be new national anticipatory institutions to help guide the evolution of this new consciousness. Hopefully, our expanded sense of identity will allow for political change and experimentation.

It is not technology per se that will do us in. In large part, it will be due to our inability to evolve from our socio-economic systems that emphasize immediate gratification and fulfillment of individuals' egos. Can we move on from zero-sum game mindset to the symbiotic mindset that actually characterizes nature? Can we deeply understand that winning a small economic or political victory today at the expense of human survival over the long term is not winning? It is the definition of losing. If we become extinct, the Dow Jones Industrial Average will be zero (0!) for evermore. The total value of the wealth of the world's nations will be zero (0!) for evermore. World domestic product will be zero (0!) for evermore. This would be a sad and total loss.

The program set out herein is expensive, well expensive given today's conventional thinking. Many already argue that we cannot afford the investments to deal with climate change, much less extinction events that may be thousands of years into the future. It will also be argued that investments in maintaining the journey will reduce funding to fight poverty and advance other social issues. Of course, these arguments are built on false assumptions and false choices. For example, human societies can decide to reduce or even eliminate defense spending. Human societies can raise taxes. Human societies admit that much of what we spend our money on is superfluous, such as the vast amounts of money we spend on knick-knacks, computer games, and advertising. Much of what we spend our money on arises from the need to ensure that nobody gets away with anything. Here I am thinking about all of the money spent on accounting systems, audits, and such. I am also thinking about the regulatory burdens placed upon companies and individuals alike that add extra costs to doing the right things because a few among us choose not to. The point is that there are resources available in today's system to take this program on if we so choose.

Embrace the Journey, Honor our Future Generations!

Epilogue

Though the contents of this book were years in the making, I began earnestly writing the book in Fall 2019 with the goal of completing this project by the fall of 2020. The months rolled by and by March 2020 the United States was engulfed in the COVID-19, coronavirus global pandemic.[1] Like many, I started working from home every day and followed the sheltering-in-place directive. As I edited the various chapters, it seemed like an excellent opportunity to use the pandemic to assess the strengths and weaknesses of the ideas presented earlier. And thus, the idea for this Epilogue was born.

Let me begin by stating that COVID-19 was a global pandemic. The virus first emerged in Wuhan, China, where it is thought to have jumped to humans at an open-air 'wet market'.[2] The outbreak was first reported at the end of 2019. Afterwards, every part of the world was hit by the coronavirus. The global health community had anticipated a global pandemic for many years, though it was not possible to anticipate the emergence of this particular virus.[3] Unfortunately, in this case, anticipation did not equal preparation, using the United States as an example.[4] It is clear that the U.S. government ignored pre-pandemic preparation efforts and advice from its own public health officials. As noted in Chapter 3, the desired time frame to have treatments and vaccines was much, much shorter than the lead times needed to develop new treatments and vaccines.

In the United States during the first phase of the pandemic, health care in hot spots such as New York City was overwhelmed with patients. The United States lacked protective gear, including N-95 masks, ventilators, and other necessary equipment. Testing kits were unavailable nationwide, so it was impossible to estimate the extent of the virus and its mortality rate. No effective treatments were available. No vaccines were available, either. Public health officials advised that people stay away from each other, wear masks, and frequently wash their hands.

Governments across the globe locked down their societies. Large public gatherings were prohibited. Business establishments such as bars, restaurants, and movie theaters were closed. The air travel and associated tourist industries saw huge decreases in revenue. As possible, office workers were told to work from home. Schools went online or were closed. Economies suffered major disruptions. Initial financial bailouts managed by the U.S. Congress and the U.S. Federal Reserve System were enormous but were essentially ad hoc and reactionary (and were also plagued by corruption). It is too early to understand what adverse unintended consequences will flow from these hasty decisions.

Fortunately, by itself, COVID-19 did not rise to the level of a human extinction event. The mortality rate from COVID-19 was high in comparison to seasonal flus but over 99% of those infected by the virus survived. Many of those infected did not exhibit any symptoms. Unfortunately, COVID-19 has caused hundreds of thousands of deaths, many of whom were young, healthy individuals before they were infected.

On the other hand, responses to the pandemic easily could have veered into a Singular Chain of Events Scenario (and still might). The 45th president of the United States (pOTUS) tried in vain to ignore the crisis at first[5] and basically mismanaged the federal reactions to the pandemic at every stage. pOTUS, seeing the economy and reelection chances crumble, pushed to reopen the economy before the crisis passed, and irresponsibly promoted not only unproven treatments but also extraordinarily dangerous treatments, such as injecting disinfectants. pOTUS picked a fight with China and refused to allow the United States to participate in international collaborations to contain the virus. If governors and mayors across the country had not issued shelter-in-place orders themselves, pOTUS could have actually pushed the world into an SCES.

As mentioned earlier, data were lacking about the extent of COVID-19 infections and the mortality rate. Due to lack of testing, quarantines were not well implemented nor was contact tracing. While there were definitely some population groups at more risk to complications from COVID-19 infections (e.g., elders and people with underlying health issues), it is still a mystery why some otherwise healthy individuals succumbed to the disease. I think that had the world had the WEMS in place, especially the component related to human exposomes, the worldwide response to COVID-19 would have been more expeditious and effective.

Will the COVID-19 pandemic lead to improved and more futures-oriented public policies? I think the answer to this is a timid yes. At least in the United States, offices to deal with future pandemics will be better funded and the United States will begin to stockpile masks, ventilators, and other supplies to use in future pandemics.

However, the more important question with respect to this book is whether humanity will become more futures-oriented with respect to perpetual obligations to future generations and supporting humanity's journey of self-actualization through time and space. As far as I can tell, the answer to this question is a weak no. The pandemic has caused people to focus on the here and now and maybe to focus on being better prepared for the next pandemic. This is the influence of personal experience.

There has been some hope that the shared global experience of the pandemic might then lead to increased global concerns about climate change, which does remain an extinction-level threat. Researchers were able to estimate reductions of emissions of GHG into the atmosphere attributable to the lockdown of economies (e.g., from fewer plane flights and automobile trips). Many commented on improvements in air quality, too. However, by and large, the needle has not moved in the United States with respect to climate change. The current administration continues to assault regulations that could mitigate the emissions of GHG. No mention has been made to reviving the U.S. economy in ways that could be considered environmentally responsible.

I have also not detected any specific references to obligations to future generations in discussions about the pandemic. Also, some important seeds may have been planted. First and foremost, the pandemic has shown that the vast majority of Americans are caring and empathetic people. Social distancing and self-sheltering were quickly adopted for self-interest and also altruistic reasons (i.e., to not contribute to the spread of the virus). The angel needed to whisper in the ears of Americans about perpetual obligations to future generations might truly exist. I am heartened that there is some evidence that people could gain psychological benefits from caring about future generations and also be positively influenced by rational arguments.

Additionally, a very limited notion of the journey has now entered the public consciousness with respect to the journey in the form of 'flattening the curve'. We all must cooperate to reduce the transmission of COVID-19 to reduce peak demands on health care and therefore save lives. Of course, flattening the curve means that society will live with the virus for a longer period of time, thus the journey characterization. Again, I think the majority of Americans understand that the journey to flatten the curve will take time and that we are all part of the journey.

Unfortunately, I need to temper this bit of optimism in three ways. First, a big question is to what degree that people in the United States and around the world trust science and scientists with respect to COVID-19. This is an important question with respect to *Perpetual Obligation #12: Fully inform current generations about futures and obligations* because anticipation is fundamentally based on science, be it climate science, social science, systems science, etc. People will not vaccinate their children to protect themselves and society against illnesses and pandemics if they do not believe in medical science. They will deny the climate crisis if they do not believe in climate science. This situation has worsened during the coronavirus pandemic as many conservative religious leaders have adopted the not very conservative opinion that the virus will not infect their faithful. It also does not help that many political leaders pander to these constituents if not also share anti-science beliefs.

On balance, I think that the majority of Americans believe in science and put much more stock in public health officials than many politicians. People are eager for science to find effective and safe treatments and a vaccine for COVID-19. The prospect that vaccinating individuals might change the essence of humanity (i.e., *Perpetual Obligation #7*) is of no concern whatsoever.

Second, it is clear that a non-negligible proportion of the American society simply does not care. Or at least their speech and behavior so suggest. They refuse to wear masks or social distance as signals of their political affiliation and as a proud demonstration of their rejection of science and notions of the public good. They continue to congregate in bars, in churches, on beaches, and at massive motorcycle rallies.[6] So, it is not a matter of whether they believe in science or not, they just don't care about public health or climate health. There is no angel whispering in their ears to care about the greater good much less the future which can be measured in weeks and months, much less years and decades. I am reminded by a book by Barbara Tuchman called *The March of Folly: From Troy to Vietnam.* This book describes that throughout human history there have been chains of self-interested, callow, and ignorant decision-making that ultimately led to disastrous outcomes.

I am not certain what they care about beyond themselves, beyond their identity as advocates of the motto 'Live free or die'. Are they simply intellectually impaired? Prone to being persuaded by conservative rhetoric? Disdain for the 'liberal elite'? What truly puzzles me is that they also do not seemingly care about their own health or the health of their children. To see a similar scenario taking shape and possibly unraveling in American society in real time is beyond shocking. It is an existential nightmare. Still, I believe that there are enough rational and empathetic people in the United States and around the world to save everyone. I just wish it wasn't so hard, since dealing with all of the threats to our existence is hard enough without ill-conceived resistance. If those who do not care come to rule the world, then the end of humanity is in sight.

Lastly, it appears that it takes a very high risk of death to move public opinion and prompt behavioral change. I think COVID-19 places *Perpetual Obligation #5. Continuously work to reduce the risk of death from involuntary environmental risks that are greater than the ethical standard* in an interesting light. Recall that the proposed threshold is that involuntary environmental risks should not exceed one in a million deaths (i.e., $10{-}^{6}$). This threshold is only ad hoc and doesn't seem to ever impinge upon the consciousness of the average American as they live their daily lives.

However, the risk of death from the coronavirus, which I believe most people believe is largely an involuntary risk, is very much in the public consciousness. For many, the risk of death is unacceptably high in spite of the fact that there are major uncertainties around what that risk is. We also know that individuals' risk perceptions do not always align with risk estimates communicated by experts due to various cognitive heuristics and biases. For example, people will tend to overestimate the risk of dying from dreaded diseases such as cancer. The media has communicated that dying from COVID-19 can be painful if not gruesome, so some people may be overestimating the probability of dying from COVID-19, while as alluded to earlier, many others may simply be in denial.

Still, let's explore some numbers. Currently, the lower probability that a person infected with COVID-19 will die seems to be about 1%. An upper probability seems to be about 5% or even up to 50% for the geriatric or infirm. A lower probability of the percentage of a population that could be infected is around 10%, with an upper probability of around 50%. Given these numbers, the lower probability of a person being infected and dying from the virus is 0.01% (i.e., 10^{-3}) and the upper probability is 2.5% (2.5 x10^{-2}). These risk estimates are considerably higher than 10^{-6}. Thus, on the one hand, there appears to be a risk threshold that can prompt major changes in human behavior. On the other, danger needs to be quite imminent to prompt behavior change, which calls into question whether humanity can be truly proactive in meeting any of its perpetual obligations when danger is mostly theoretical, in the distant future (defined to be anything beyond a few years!) and the risk threshold is much, much more stringent.

It is interesting that some people have used terminology found in the action framework presented in Chapter 7 to characterize the level of societal mobilization needed to deal with COVID-19. Specifically, the terms Manhattan Project and War Footing have been used. Though pOTUS has referred to himself as a wartime president, the United States has not mobilized its manufacturing infrastructure to deal with the crisis. Money has not been diverted from the Department of Defense to fight the epidemic. In fact, recent reports suggest the reverse has happened. No attempts have been made to fashion an economy better suited to fight this war or wars against future pandemics, in fact the opposite has been advocated.

A really interesting thought experiment is how would the United States' and world's policy actions been different with respect to COVID-19 had a fully operational and effective GADDS been in existence and effectively used. From the perspective of the United States, the system could have better convinced states and cities to work in concert to lock down hot spots and implement nationwide shelter in place plans. Instead, haphazard responses allowed the virus to circulate unabated through large swaths of the country to then return to original hot spots in the United States. Ideally, GADSS could have allowed expeditious international cooperation with respect to international travel and trade and could have promoted transparency in tracking and reporting on the disease. These data could have been extraordinarily valuable to scientists trying to understand COVID-19's transmission and mortality rates. International cooperation might have also allowed more effective plans to manufacture everything from masks to ventilators. Conversely, users could have used GADSS to game the impacts upon their constituents of irresponsible actors.

Another question worth pondering is whether the existence of NAIs around the world would have helped prevent or at least ameliorate the COVID-19 pandemic. More specific to the United States, would the existence of a prominent and well-respected Court of Generations have made a difference? I would love to say, yes, absolutely, but the answer is probably no, at least in the short term. If painful personal experience is needed to change behaviors and policies even as the news is filled with scenes of death and destruction from just down the road, then admonishments of current generations by the COG to act on theoretically sound arguments probably would have gone unheeded, though not necessarily completely ignored. However, let's assume that the COG did emphasize risks from pandemics and now those risks were realized. Let's also for the sake of argument assume that eventually every community in the United States was hit hard by the coronavirus. Then painful personal experiences might be able to give the COG more credibility and could then increase its influence on near-term public policy with respect to the gamut of existential risks and scenarios. I can imagine similar stories might pertain to other countries.

Cynically, one can also ask how many NAIs could be free of corruption, free to meet their responsibilities with integrity, transparency, and accountability? How many would have the resources to meet their responsibilities effectively? It is quite arguable that the current pOTUS would have worked to corrupt the U.S. COG, deny it funding, and otherwise tried to undermine its operation for political gain. It is also quite arguable that keeping NAIs out of everyday government and away from policy making and funding decisions seems like a good strategy to prevent corruption of the NAIs. Also, being optimistic, if a few NAIs can become established and operate as expected, then more NAIs would be established around the world and aspire to meet high standards. Sometimes, it is easy to become impatient, even as a futurist!

I think that the very quick cratering of the United States and other economies might also plant a seed for the establishment of perpetual organizations, such as those presented in Chapter 8 to deal with nuclear wastes and archiving human knowledge. At this point it is easy to imagine funding being cut for such organizations, which then might cripple if not dispatch their organizations into oblivion. However, this would not be an issue if they were self-sufficient.

Chapter 9 addresses changes to societies needed to ensure the journey and meet perpetual obligations that could be described as passing through a Singularity. Experiences with the COVID-19 pandemic suggest that humanity is very far away from the Singularity and achieving required changes will be exceedingly difficult.

For example, it is frequently argued that democracies are a boon to humankind, that the world will be better if all countries were democratic, notwithstanding Winston Churchill's view of democracy. However, for me anyway, the pandemic has exposed a somber flaw in the democratic system: there is no guarantee that elected officials will possess any semblance of leadership capability to effectively manage crises, or that they will rely on evidence in decision-making. Of course, it was very naïve of me to believe otherwise, but I guess I also needed the painful personal experience of witnessing the failure of the current president to exhibit any scintilla of leadership capability. This U.S. president has so far shown no ability to grow into the job, while many U.S. state governors showed leadership in their absence.

Two important aspects of leadership are empathy and intelligence. It is now clear that many Republicans in the United States, including the president, can only see people as cogs in the industrial machine, cogs that ought not be overpaid, cogs that should be blamed for their low incomes, unemployment, and lack of health care. They are actively working to hide the fact that the system works against the average working American and not for them. A politician in Texas actually said that older people (i.e., not working and contributing to GDP) should be prepared to die of COVID for the sake of the economy. This statement is a perfect example of the inverse of what sustainability and concern for current and future generations is all about.

In retrospect, pOTUS was elected not to be competent, not even to be empathetic, but simply as a tool in a zero-sum game. One view of pOTUS is as an anti-futurist. Their brand of populism references an imaginary past where life was simpler and homogeneous culturally. The future only holds threatening change, so let's not only stop change but also move backwards in time. This worldview is a challenge to futurists and a menace to all of the endeavors supported herein.

Here is a second view. When I was young, I read Isaac Asimov's Foundation trilogy. It has been a long time and my memory is really quite hazy, but I do remember Hari Seldon and psychohistory. And I remember the character known as The Mule. The Mule represents an unpredictable and destructive force unleashed upon psychohistory mainly. The Mule is driven by a predatory narcissistic personality. It possesses a mutant ability to influence the thoughts of others. Personal power and revenge reign supreme. The public good is not a recognizable concept to The Mule. Interestingly, The Mule is depicted visually as a clown. And, yes, I am not the first to find similarities between The Mule and pOTUS.[7]

Can humanity survive pOTUS The Mule and other Mules that one can anticipate will emerge over time? Might Mules be endemic across the universe and be the main explanation of Fermi's Paradox? In Asimov's trilogy, the Second Foundation was able to defeat The Mule, but not without much pain and suffering. Maybe this is the optimistic moral of the trilogy. Nothing will be easy for humanity. Our journey of self-actualization and progress toward meeting our perpetual obligations will not be linear or steady. It will be non-linear. It will be filled with mistakes and self-inflicted pain and suffering. Hopefully, there will be enough angels in enough ears of people through time and space to ensure that humanity can always recover and prosper.

As I said in the introduction, most of my remarks focus on my experience studying the U.S. response, which has led to more COVID cases and deaths, so far, than any other country. However, looking across the globe, the United States was not the only country whose governments did not respond well. The failure of China to be open and accountable early on greatly worsened the pandemic. Totalitarian regimes in Iran and Russia ultimately failed to protect the health of their citizens. Politicians in many democratic countries, such as Italy, were also slow to respond. At least the freedom of the press in the democratic countries (albeit to the chagrin of the U.S. president) was able to mobilize society and hold politicians somewhat responsible. This all being said, the case can still be confidently made that very long-term experimentation with new forms of political systems seems to be a good idea for humanity to embrace.

The second topic addressed in Chapter 9 was economics. It was argued that many aspects of capitalistic systems and modern market economics are not in alignment with sustainable production systems or basic futures thinking. The devastation of national economies around the world highlighted just how fragile our economic systems actually are. Who could imagine that unemployment in the United States would change from a generational low to exceeding the levels experienced during the Great Depression in the 1930s in a matter of two months. This dramatic decline in non-essential employment brings up the question for me as to what the global economy is all about. Are all of the jobs in service sector that have been put on hold or permanently lost truly important, nice to have to improve our quality of lives, or, in some sense, created by consumer-oriented economic policies to ensure that people have some type of employment and guarantee that wealth will be continued to be created for the owners of capital?

It can be argued that the global economy does (or did) a decent job of producing 'stuff' but that it is incapable of quickly shifting to producing more of what is needed (e.g., masks, ventilators) and ensuring that these products are equitably distributed. In the United States, many companies that remained open for business did a very poor job of protecting their workers (e.g., meat-processing facilities). It is an open question about whether these companies were unable or simply unwilling to adjust processes. Small businesses were especially hard hit in the United States. As I write this, the conventional wisdom is that thousands of shuttered small businesses will not survive the pandemic.[8]

One of the ideas presented in Chapter 9 regarding economics centered on increased self-sufficiency in production. I can imagine that the U.S.

economy would be in a much better place if most households were able to produce most of their own energy and food, and manufacture everything from clothes to pharmaceuticals. I can imagine vibrant neighborhood economies that could effectively weather this pandemic and others to come with less economic disruption. It will be interesting to see if there is an increase in self-sufficiency over the coming years.

It has been very fascinating to see how American culture has reacted to the pandemic and lockdowns. On the whole, most of society has adjusted. Sheltering in place has led to much creativity in how to pass the time. The videoconferencing platform, Zoom, has led to a new verb, to Zoom, much like Google is also now a verb, to Google. I have seen some very entertaining live media content on Zoom and YouTube. The episode of Some Good News with John Krasinski with the entire cast of Hamilton on screen performing simultaneously was particularly inspiring. I think my point here is that culture can adapt creatively, which is good news moving into the future.

I think that the lockdowns have had an interesting range of impacts on people's self-identity. In my case, my perspectives have refocused on family and friends. Work is still important. I still put quality time in on this book while sheltering at home (hopefully a good use of my time and your time, the reader!). But, the need to travel to a professional meeting to make this or that presentation has almost completely dissipated. I miss being in the office, but my role right now is to stay healthy and safe and not to put others' health and safety at risk. I am quite aware that I am privileged to be able to stay at home when others need to be out in the workplaces. I am truly in debt to those in health care working with COVID patients and those in meatpacking plants and elsewhere helping to make sure our food supply is stable and safe.

The limits of adaptive work were clearly on display during the epidemic. It is absolutely clear beyond a shadow of a doubt that any and all attempts to engage pOTUS during this time period in a productive adaptive work process would have completely failed. It is also clear that adaptive work efforts would have been rejected out of hand by millions of others in the United States. Is adaptive work a fool's errand?

Addendum to the epilogue: Black Lives Matter

In May 2020, protests erupted throughout the United States against racism and police brutality. The precipitating event was the killing of George Floyd by a white Minneapolis policeman on May 25, 2020. Mr. Floyd was one of thousands of black Americans killed by police in recent years. His death was captured on video, which quickly went viral. His death also coincided with COVID-19 lockdowns, which enabled tens of millions of Americans to have the time for reflection, empathy, anger, and action.

Many of the protests have been couched in terms of support of the Black Lives Matter (BLM) movement. The protests have been mostly peaceful, though interventions by white supremacy groups, egged on by pOTUS, have resulted in violence and death. American protesters have been joined by other concerned citizens around the world. The massive protests also brought people together, probably way too close together according to public health officials. However, many of these officials agreed that the protests were worth the risk of higher rates of COVID-19 infections.

Slaves were first brought to America in 1619, just over 400 years ago. Slavery was abolished in the United States in 1865. One hundred years later, the Voting Rights Act of 1965 was signed into law in the United States. Has any progress been made against racism and civil rights in the United States in the past 400 years? Yes. But, not enough, nowhere near enough. Racism has not gone away and is still a stain and danger. Is this a watershed moment in American society? Maybe, today it feels like it, but tomorrow?

I think we are witnessing the non-linearity of progress with respect to ending racism, and the pain of non-linearity. We are also attesting to the timescales necessary to make social change. The American history of oppression of African Americans now spans 400 years. Slavery itself dates back thousands of years, if not deep into human pre-history. We ought not wait another day to work toward rectifying the wrongs of racism. Not another day. Yet, from a futures perspective, seeing how hard change is with respect to racism, setting a goal of passing through the socio-cultural Singularity by the Year 3000, one thousand years from now, seems to be a bit of a rush. Unfortunately, not meeting this goal poses unacceptable risks of catastrophe and human extinction, in my opinion.

The last point I wish to make relates to a necessary condition for change with respect to both futures and BLM. The inspiration for this concluding observation is the American professional sports establishment. Many sports leagues were anxious to resume their games. Foremost among them was the National Basketball Association, where about three-quarters of the players are black. As preparations were being made to resume the season in a bubble at Disney World in Orlando, Florida, a debate ensued about whether the restart of the National Basketball Association's season would help or hurt the Black Lives Matter movement. The central point of the debate was whether professional basketball would divert time and attention from the movement. Would 'normalization' of sports just set the BLM movement and efforts to eliminate racism back to the unacceptable prior situation?

The bigger question is whether renormalization of American society would quash our most recent steps toward racial justice. Normal life in the United States does not allow much if any time for reflection or active citizen participation. Work dominates. Many have to work two to three jobs to make ends meet. In the name of economic efficiency, production is now 24/7. Those working rotating hours are often sleep deprived and do not have a regular schedule around which to plan citizenship activities. Social capital has declined tremendously. Frequently, volunteer activities are confronted by full-time corporate representatives, with one result is that the former simply burns out. It is almost as if the economic system is designed to limit meaningful public participation.

This observation casts a very dark shadow on the ability of American society to become more futures-oriented. What could possibly change if no one has the time or energy or social capital to reflect and become involved? How could adaptive work processes make any progress if no one has spare cognitive capacity to engage in adaptive work? Could the world go on lockdown for a week or two every few years to engage in futures-thinking? Should the world regularly take time off to reflect upon and renew efforts to meet perpetual obligations, to reinvigorate the futures-angels of their souls? I think so.

Notes

1 The last edit of this Epilogue was made on September 27, 2020. According to Johns Hopkins University, as of this day, there have been 32,881,747 confirmed cases of the Coronavirus worldwide, with 994,821 deaths https://coronavirus.jhu.edu/map.html. These numbers are certainly underestimates because of lack of testing for the virus in countries like the United States and premeditated lack of transparency and accountability from countries like China about how the virus is impacting many countries around the world. The estimates for the United States are 7,079,909 cases and 204,503 deaths.
2 WebMD. 2020. Retrieved from www.webmd.com/lung/coronavirus-history#:~:text=Experts%20say%20SARS%2DCoV,"wet%20markets
3 M. Osterholm and M. Olshaker. 2020. Chronicle of a Pandemic Foretold. *Foreign Affairs*, 99, 4, 10–25.
4 I tried to keep up with developments across the world but will, for the most part, keep my comments limited to my experiences of the United States in this piece. As it turns out, the reasons for the United States' exceptionally disastrous experience provide much fodder with which to judge the potential usefulness and efficacy of many of the ideas presented in this book.
5 A new book by Robert Woodward, titled *Rage* (Simon & Schuster, New York, 2020), now confirms that this was quite intentional on the part of pOTUS.
6 Please see reports that the Sturgis motorcycle rally in South Dakota, which drew approximately 450,000 participants, has now been linked to 260,000 COVID-19 cases and $12.2 billion in additional health care costs. www.usatoday.com/story/news/nation/2020/09/08/study-260-000-coronavirus-cases-likely-tied-sturgis-rally/5750587002/
7 https://mashable.com/2017/03/15/foundation-mule/
8 I wish to thank Dave Feldman for his contributions to this part of the Epilogue.

Index

Note: Page numbers in *italics* indicate a figure and page numbers in **bold** indicate a table on the corresponding page. Page numbers followed by 'n' indicate a note.

For Product Safety Concerns and Information please contact our EU
representative GPSR@taylorandfrancis.com
Taylor & Francis Verlag GmbH, Kaufingerstraße 24, 80331 München, Germany

www.ingramcontent.com/pod-product-compliance
Ingram Content Group UK Ltd.
Pitfield, Milton Keynes, MK11 3LW, UK
UKHW020142200926
475702UK00005B/867

* 9 7 8 0 3 6 7 4 2 9 1 3 3 *